IUDA's

CBEST
MATHEMATIC

Learn From Real Examples

Study Differently

CBEST Math Extensive Analysis & Reviews.

CBEST Math Section:
California Basic Educational Skills Test

Author

Rahma, M., M.A. Edu (APU.Edu), MBA ip (WGU.Edu)
Independent Universal Development Academy (IUDA)
www.professoruda.weebly.com

Accompanied Web Presence
http://cbest-success.weebly.com

Copyright © 2014 professoruda@gmail.com

All rights reserved.

ISBN-10: 1495352722

ISBN-13: 978-1495352720

DEDICATION

People who have been contributing & sharing their lifelong experiential knowledge for the benefits of others' lives. It's dedicated for them-educators; teachers, writers.

CBEST Math Section:
California Basic Educational Skills Test

ACKNOWLEDGMENTS

Thanks to the following websites and publishers who have contributed their various thought, ideas, math articles, & CBEST course outline, NES publication, practice math test module, sample & simulation exercises in which makes me motivated to introduced CBEST math study guide.

References:

http://www.ctcexams.nesinc.com/PDF/CBEST_OPT_Math.pdf

http://www.khanacademy.org/

http://en.wikipedia.org/wiki/Mathematics

CBEST Math Section:
California Basic Educational Skills Test

TABLE CONTENTS

	Listing Preview Topics	Page
1	Preface	9
2	Introduction CBEST Mathematics	11
3	Basic Math Skills	13
4	Test Taking Tips & Directions	14
5	Math Words' Jargon	15
5	Identified Key words/phrases breakdown root meaning	16
6	Recognize Math Formulas	17
7	Identify needed information to solve the given problem	18
8	Proofread Math Backward	19

CBEST Math Section:
California Basic Educational Skills Test

COURSE OUTLINE

	Syllabus Contents	Page
1	Arithmetic Contents	23
	Arithmetic Preview	29
	Chapter One	35
2	Basic Algebra	25
	Basic Algebra Preview	31
		37
3	Geometry	27
	Geometry Preview	32
		38

CBEST Math Section:
California Basic Educational Skills Test

This Page Internationally open

CBEST Math Section:
California Basic Educational Skills Test

PREFACE

The CBEST math syllabus contains elementary through K12 math learning lesson. Passing a CBEST math test is very easy when a test-taker regularly study math. This study guide's "Learn From Examples" help them how to pass CBEST math test.

The "Learn From Example" approach gives test-takers a depth understanding, which makes them stamina, energy and confidence to overcome math puzzle things because it focuses on extensive math review. This study guide is only ever written that gives a test-taker 100% unconditional satisfaction CBEST-success.

This study materials introduced all essential math's learning techniques and strategies. Its learning approach keeps away your anxieties, frustrations, and develop math skills gradually. By the way, the CBEST test comprises of three modules; Mathematic, Reading, Writing. This study guide is one of them is mathematic.

What are the topics in this study guide covers:
- It covers CBEST mathematics: arithmetic, basic algebra, and geometry.

CBEST Math Section:
California Basic Educational Skills Test

- It covers CBEST math review in-depth explanations.
- It covers practice test explanations and analyzes.
- It covers basic mathematics formulas, methods, review procedures.
- It covers mathematic terminology and concepts
- It covers arithmetic concept and various examples
- It covers basic algebra equation and solving technique
- It covers geometric various diagram and their application
- It covers measurement various units that we use everywhere
- It covers graphs, charts, and graphs' representation and concepts.
- It covers test taker strategies, techniques, and test preparation and precaution.

This CBEST math study guide is written to aim pass CBEST math exams, but not substitute entire courses. In fact that if a test-taker study this guide regularly then s/he shall have a good chance to get the CBEST math done.

CBEST Math Section:
California Basic Educational Skills Test

Introduction CBEST:

CBEST stands for California Basic Skills Test

CBEST basic skills is required those who interested to be a substitute teacher in kindergarten through 12th grade public school system in California. The CBEST test is conducted by the California Commission on Teacher Credentialing (CTC), www.ctc.ca.gov.

The CBEST math measures general areas; arithmetic, algebra, geometry. Teachers required skills taught to students that essential to both in the traditional classroom learning and non-traditional learning approach.

CBEST comprises of three modules: English Reading comprehension, research, critical analysis and evolution; English Writing two essays; Basic mathematic. These examinations have a 4-hour test session both PBT or CBT. You may take whole 4-hour for one module or two, or all of them together. Although any test taker may divided 4 hours' time among the 3 tests, in order to make sure enough time to complete all three module.

The CBEST math test has 50 multiple questions along with 5 choices. Passing score calculation, each section raw score range starts 20 to 80. It then converted into passing or failing criteria. To achieve passing level performance, it requires 35 + questions right out of 50. When a test-taker get passing criteria, it would reflect a score between 37 to 41. It is the minimum passing performance level ball-park range 70% +. However, a test-taker must achieved minimum 123 to pass all 3 sections.

References:

    http://www.ctcexams.nesinc.com/register.asp

    http://career.ucsd.edu/_files/tchgk-12.pdf

Basic Math Skills

Basic math skills, the CBEST questions selection comes from arithmetic, fundamental concept of algebra, reading interpretation of graphs, tables, charts, measurement unit, reading ability of fraction, decimal, ratios, percentage, and probability, statistical survey results and their interpretation.

Test taker must have adequate demonstration ability to using numerous math formulas; equations, probability, and other reasoning knowledge, which curricula ranging from elementary grades to K12 and beyond.

CBEST Math Section:
California Basic Educational Skills Test

Test Taking Techniques & Strategies

CBEST Math test has no penalty for wrong answer. Therefore, it's recommended to eliminate a choice one after one and finally chose a best guess that you think best possible answer. For instance

Which of the following number is in the tenth digit place value?

    A.   10.0
    B.   1.0
    C.   0.1
    D.   0.01
    E.   0.0001

First, A. 10 is ten. It is not tenth thus eliminate it. Second, B. is 1.0, Choice D. is hundredth (100th) and E. are out of question thus C is the best guessing response. Therefore, it's wise decision to select a best guessing choice that corresponding to the numbers on the question-sheet through eliminating process.

CBEST Math Section:
California Basic Educational Skills Test

Math Words' JARGON:

There are few word jargon questions are in the CBEST test. A test-taker can easily overcome and solve such type of CBEST math if s/he reads the problem's wording/phrasing.

Wording/phrasing deals different approaches with a single number. For example, round-off, nearest number or nearest hundred, tens place, tenth place, hundred place, hundredth place, thousandth and so on. To deal with wording/phrasing, test-taker needs to become familiar with them before take a schedule CBEST test date.

Identify Keywords/Phrases & Breakdown Their Root Meaning:

Pull-out word(s), phrase(s), and facts that related with the problem, which helps but remember that not always work out.

In mathematics, for example, than, is, less are respectively represents "plus + or more also + , equal to = , and minus sign - . Without knowing them, a test-taker can't pass the CBEST math test. More such examples are furnished in this study guide.

Recognize Math Formula

Math formulas are essential learning elements to every student. To solve math problems, for example, 0.666 is equal to 0.67 (round-off), it then becomes 67% or 67/100. These all numbers represent respectively decimal (0.67), percent (67%), and fraction (67/100) are same value but different representation/terms. They come from math formula. Aren't they? Yes, they're. Therefore, a test-taker must identify such relationship immediately upon see them or able to understand and transformation techniques among them, which helps test-taker to solve problem work forward.

CBEST Math Section:
California Basic Educational Skills Test

Identify Needed Formula(s)

How Formula Works!

To solve a problem, some time you do not have enough information or insufficient data or problem did not provided needed data.

What shall you do?

Well, you can solve it if you remember or focus on several formulas. One of the road maps is the missing number.

What is the average of the following equation?

5+5+5+5= missing

Average = ?/?

What is the average number?

This equation needs to find out the total of "missing" i.e. 20.

As we know the average formula; total divided by number of items of that equation has holds i.e. 4 items.

As we know that average or math's mean formula is the sums of data (5+5+5+5 =20) divided by the number of items (1+1+1+1=4, it means 5 appeared 4 times) in the data will give the mean average.

Replace them; average = ?/? = 20/4=5

An average number is not given but it can be solved when each item's costs are known separately.

CBEST Math Section:
California Basic Educational Skills Test

To adding up all item number then divided accumulated number by the total items. It gives you the average or mean results.

Find Answer Backward:

Work backward from the answer choices. It very simple task, for example,

What is value of **Z**?

When,

$z/2 + \frac{1}{4} = 1$

    A. 1

    B. 1.5

    C. 2

    D. 2.5

    E. 3.0

Work backward, plugging- in each choice with the equation.

CBEST Math Section:
California Basic Educational Skills Test

Glance of
CBEST Math Topics

CBEST Math Section:
California Basic Educational Skills Test

Arithmethic Contents	Page.
Introduction	25
Basic Math's Symbols & Terms	26
Arithmetic Order of Operation Preview	30
Addition ------------------------------------- 30	
Subtraction --------------------------------- 32	
Parentheses -------------------------------- 35	
Parentheses in Algebra -------------------- 37	
Parentheses order of operation ---------- 39	
Order of Operation Methodology ------ 43	
Multiplication ------------------------------ 45	
Division ------------------------------------- 47	
Understand Positive & Negative Scale Reaing:	50
Ruler/Scale Measurement ----------------	
Analyze Scale Elements ------------------ 50	
Math Properties:	51
Addition Axioms ------------------------------------**51**	
*Commutative	
*Associative	
*Additive Inverse	53
*Distributive Property	

CBEST Math Section:
California Basic Educational Skills Test

Multiplication Axioms* ------------------------53 Distribution Axioms* -------------------------53	
Fundamental Statistics: 　　Arithmetic Mean/Average -------------- 55 　　Mode--------------------------------------60 　　Median------------------------------------ 61 　　Range ------------------------------------ -63	56
Arithmetic Numerical number: 　　Number Place Value ----------------- 64 　　Place Value Description -------------- 69 　　Whole number --------------------64/ 109 　　Fractions ----------------------------76/96 　　Decimal ------------------------------- 101 　　Decimal Fraction --------------------- 98 　　Percent --------------------------------- 115 　　Square -----------------------------------79 　　Exponent -------------------------------80 　　Mixed Numbers ---------------------- 82 　　Adding,Subtraction Mixed Number 84	64
Rounding Off Concept of Numbers	67
Rounding Off Methodology	68

CBEST Math Section:
California Basic Educational Skills Test

	Cost Estimating Method	70-76
	Sum, Difference, Multiply, Quotion etc	
	Decimal Digit Measuring/Counting	105
	Tenth, Hundredths, Thousandths --105	
	Adding, Subtracting, Multiplying, Dividing ---------------------------------110	
	Changing % to Decimal Point ----- 115	
	Arithmetic Word into Sign --------- 134	
	Percentage Change Formula ------- 138	
	Prime/Non-Prime Number ------------ 140	
	Ratio & Proportion	139
	Prime/NonPrime	142
	Factoring	145
	Length/Weight Measurement Units	150
	Length Traditional Vs Metric ------150	
	CBEST Watch ----------------------- 160	
	Weight Measurement -------------- 168	
	CBEST Watch ----------------------- 169	
	Weight Units ------------------------ 170	
	Liquid Measurement --------------- 170	
	CGS System ------------------------- 172	

CBEST Math Section:
California Basic Educational Skills Test

	CBEST Watch ---------------------- 172	
	Basic Probability Principles:	181
	Probability --------------------------------181	
	Predicted outcomes ----------------------185	
	Permutation -------------------------------187	

CBEST Math Section:
California Basic Educational Skills Test

Arithmetic

Introduction

Arithmetic came from Greek word known as *'Arithmos'*. it means number, which is elementary level of mathematics. The CBEST arithmetic covers very popular tasks ranging from simple day-to-day business counting calculations such invoice total amount that comprises of a dollar fraction of a quarter, dime, nickel, penny etc.

Arithmetic refers to the simpler form of properties that using the traditional operations of addition (+), subtraction (-), multiplication (x), and division (/) with smaller values of numbers.

CBEST Math Section:
California Basic Educational Skills Test

Basic Math's Symbols & Terms:

1. Equal to sign =
2. Not Equal to ≠
3. Greater Than >
4. Less Than < (opposite sign of greater than)
5. Greater Than or Equal ≥
6. Equal to or Less than ≤
7. Parallel to | |
8. Perpendicular sign ⊥
9. Natural Numbers – counting numbers 1, 2, 3
10. Whole Numbers – counting numbers starting with 0,1,2,3...
11. Integers – negative, positive, whole numbers, and 0...-4, -3, -2, -1, 0, 1, 2, 3, 4
12. Even numbers 0, 2, 4, 6 ... (divided by 2)
13. Odd numbers 1, 3, 5, 7, 9.. (Not divided by 2).
14. A composite number is any number, greater than 1, that is not prime number. The composite

CBEST Math Section:
California Basic Educational Skills Test

number that divisible by more than 1 and itself such as 4, 6, 8, 9, 10, 12, 14, 15...

15. First ten prime numbers are listed herein. They all, prime numbers, are divisible by itself and by 1 (one). 2, 3, 5, 7, 11, 13, 17, 19, 23, 29. Example, 2/2, 2/1, 3/3, 3/1 and so on.

16. Zero and 1 are not prime numbers

17. Except 0 and 1, a number is either a prime number or composite number.

18. Squared: when numbers are multiplied by themselves. For instance, (12 x 12) =144.

19. Cubed: When numbers are multiplied by themselves twice, for instance, (2x2x2) =8.

CBEST Math Section:
California Basic Educational Skills Test

Order of Operation Preview:

P E D M A

First Solve Parentheses: P

Second Solve Exponents: E

Third Solve Multiplication (M) and Division (D) from left to right: MD

Forth Solve Addition (A) and Subtraction (S) from left to right: AS

An Example of PEDMA solution:

$6 + (6 * 5^2 + 3) = ?$ Solve all the component inside parentheses

$6 + (6 * 5.5 + 3) = ?$

$6 + (6 * 25 + 3) = ?$

$6 + (150 + 3) = ?$

$6 + (153) = ?$

$6 + 153 = ?$

$= 159$

CBEST Math Section:
California Basic Educational Skills Test

A pair of ( ), Power, Multiplication *, Division /, Addition+, Subtraction –

Explanations:

1. Parentheses: date located inside the parentheses first. It means all parentheses pair data have to be solved.

2. Powers: Power operation like 2^2, 2^3 etc.

3. Multiplication operations.

4. Division operation.

5. Addition operation.

6. Subtraction operation

Interpretation of Math Words

Math words and phrases are key concept such as addition/add, subtract, multiply, and division. For example, add or/and multiply gives understand the relationship between price and multiple units' price total. The CBEST skill development needs the following terms to solve many of its math problems

1. **Addition**:

Addition, sum, total, plus, increase, more than, greater than, etc. are substituted addition.

Addition means plus (+) Sign:

Both numbers are positive (+).

When adding two positive numbers with the positive (also known as plus or + sign) then add the both sides + sign. For example,

$$\begin{array}{r} +9 \\ +4 \\ \hline +13 \end{array}$$

CBEST Math Section:
California Basic Educational Skills Test

Note: Theoretically we know that both sides need + sign; however, it's not norm to usage + sign as it illustrated above.

Practically it (+) usages;

$$\begin{array}{r} 9 \\ \underline{4} \\ 13 \end{array}$$

(without sign (+) mark it's norm that they are positive numbers.

Adding two minus value/numbers:

When adding two minus (-) or negative numbers, which become plus (+) sign. It shown like this: {(-) + (-)} = +.

For example,

$$\begin{array}{r} -7 \\ \underline{+\ -3} \\ -10 \end{array}$$

Adding two negative values, (-7)+(-3), which becomes -10

CBEST Math Section:
California Basic Educational Skills Test

Adding positive and negative two numbers:

When adding two numbers that have one + sign number value and another – sign or negative value, which results two values' different. The result can be either positive value or negative value.

When positive number is larger number than negative number, which subtracting results will be positive (+) sign number.

When negative number is larger than positive e number, which subtracting results will be negative (-) sign number.

Look at the following examples, which sign get precedence.

For example,

```
  +50         -50
+ -70       + +70
  ---         ---
  -20         +30
```

```
For example,

      +500              -500
      -200              +200
      ----              ----
      +300              -300
```

CBEST Math Section:
California Basic Educational Skills Test

Note: Two colors Blue and Red respectively representing positive number and negative number. Color uses for readability purposes.

Subtract:

Subtract word substituted many different ways:

Difference, reduced, less, decrease, minus, fewer, remainder, and have left.

Subtracting larger positive (+) Number:

Subtracting a smaller negative (-) number from a larger positive (+) number, this operation outcome is always positive (+) value thus arithmetic usages it positive (+) sign symbol. However, if positive number is location at the begging of an equation then + sign is not required such as 5+5=10.

First number without + sign assumes positive value like 5 and next 10.

Another example is, 6-2=4. 6 is positive and next total 4 is also + thus none of them uses + sign in from of them.

CBEST Math Section:
California Basic Educational Skills Test

Subtracting larger negative (-) Number than positive (+) number:

Subtracting a larger negative (-) number from a smaller positive (+) number. This operation outcome is always negative (-) value. Therefore, arithmetic usages it negative (-) sign symbol. However, put negative (-) sign in front of the subtracting number value is necessary. Indeed, without minus (-) sign, negative value could not be identified. look at the following exercise provided hereafter.

For example,

Operation of a **larger positive (+13) number** subtract a negative (-7) number, which becomes +5. It illustrated the following way:

```
 +13 -->    +13
 -7  -->  + -7
           ─────
            +5
```

CBEST Math Section:
California Basic Educational Skills Test

Another example of two negative numbers' subtraction.

```
For example,
   -50   -->       -50
   -70   -->   +   -70
                  -120
```

Note: The operation;

(-50) + (-70) = + (-120) = -120

It demonstrated two negative numbers are added, but their sum still negative values. Indeed, they are addition value of two negative sums.

CBEST Math Section:
California Basic Educational Skills Test

QUESTION #8: Multiplication and division

> 8. Rob uses 1 box of cat food every 5 days to feed his cats. Approximately how many boxes of cat food does he use per month?
>
>   A. 2 boxes
>
>   B. 4 boxes
>
>   C. 5 boxes
>
>   D. 6 boxes
>
>   E. 7 boxes

1 Month = 30 Days

Rob needs 1 box of cat food every 5 days.

So, 30 divided by 5 equal to 6 boxes

The choice is **D**.

Parentheses ( ):

Parentheses are a pair of symbol look like this: (). A pair of parentheses (), which have no value; however, they assist to organize mathematics calculation. Besides parentheses, two others parentheses alike symbols are also used in math they known as bracket signs. They are called curly bracket {} and square bracket []. In math, these three types symbol encloses numbers, or figures, and words so as to separate context.

Parentheses enclose numbers for calculation shown below:

(5+2), (5-3), (5x3), ($\frac{6}{2}$)

Parentheses hold a pair or group of numbers. In math solution, everything inside parentheses must be solved first prior others. Operation of order in math, parentheses operation has the first precedence than any other operation.

For example:

CBEST Math Section:
California Basic Educational Skills Test

$(5+2) + (5-3) - (5 \times 3) \times \left(\frac{6}{2}\right)$

$= (+7) + (+2) - (15) \times \left(\frac{3}{1}\right)$ Parentheses first

$= (+7) + (+2) - (15) \times (3)$ next multiplication

$= +7 + 2 - (15 \times 3)$

$= +7 + 2 - 45$

Similar terms $(+7+2=9) = 9 - 45$

Subtraction (-) or addition (+) whatever comes first.

$= -36$

Parentheses usages in Algebra:

Parentheses are also widely used when we doing algebraic calculation.

For example,

Solve $x = ?$

$3 - (-3 + x - 3x + 1) = 10$

Or $3 - (-3 + (x - 3x) + 1) = 10$

(Similar term are x & 3x) = 10

Or $3 - (-3 + (-2x) + 1) = 10$

Or $3 - (-3 - 2x + 1) = 10$

(Similar terms -3, +1)

CBEST Math Section:
California Basic Educational Skills Test

Or $3 - (-2 - 2x) = 10$

(Precedence multiplying)

Or $3 - (-2) + 3 - (-2x) = 10$

Or $3 + 2 + 3 + 2x = 0.$

(Similar terms)

Or $8 + 2x = 10$

Or $2(4 + x) = 10$

Or $8 + 2x = 10$

    Or $2x = 10 - 8$

    Or $2x = 2$

    Or $2x = 2$

    Or $x = 1$

Parentheses are mostly useful in math, it uses all of the operation of the following signs; +, -, x, /. However, parentheses along with multiplication has the first precedence than other three (+, -, /) signs noted above.

Sign/Symbol legend:

Add sign +, Subtraction - , Division / & multiplication symbol X

CBEST Math Section:
California Basic Educational Skills Test

Parentheses Order of Operations:

When one problem contains all signs such as addition, subtraction, multiplication, division, and powers then its order of operations is parentheses. Parentheses help multistep problem. It provided step by step explanations and solution tips and techniques based on order of operation theory. It also addresses multistep solution that uses parentheses usages.

Parentheses Operation Usages.

Parentheses operation has the first precedencies in an equation. Like equation

2+(3-4)=?

2+ (+3-4) = 2+ (-1) = 2-1=1,

(3-4), which needs solve first because that pair is located inside the a parentheses. All math equation has parentheses operation precedence.

Solve the equation

$$15 - 3 \times 5 + \frac{1}{2} - 7^2 + 4(3-1)$$

$$= +15 - 3 \times 5 + \frac{1}{2} - 7^2 + 4(2)$$

CBEST Math Section:
California Basic Educational Skills Test

Step #1:

Let solve the following equation.

$$= 15 - 3 \times 5 + \frac{1}{2} - 7^2 + 8$$

Continuing step #1 parentheses operation:

The above operation is continuing... 4(2) = 4x2 = 8. (2)sign falls into parentheses operation.

$$= +15 - 3 \times 5 + \frac{1}{2} - 7^2 + 8$$

$$= +15 - 3 \times 5 + \frac{1}{2} - (7 \times 7) + 8$$

$$= +15 - 3 \times 5 + \frac{1}{2} - (49) + 8$$

Step #2: Powers or square roots

Next, powers and/or squared-root function has to be solved i.e. (7x7) = 49

Step #3: Multiplication or/and division

$$= +15 - 3 \times 5 + \frac{1}{2} - (49) + 8$$

CBEST Math Section:
California Basic Educational Skills Test

Multiplication -3 x 5

$$= +15 - (3 \times 5) + \frac{1}{2} - (49) + 8$$

Let identify multiplication or/and division pairs, if exist then solve that operations.

Since, there are two places; they are 2x5=10 and $\frac{1}{2} = 0.5$

$$= +15 - (15) + .05 - (49) + 8$$

Step #4: Addition or/and subtraction.

$$= +15 - 15 + .05 - 49 + 8$$

Organize same terms

$$= (+15 + 0.5 + 8) - 15 - 49$$

Similar terms operation: Let identify all addition sign and all negative sign numbers then added up them separately.

So, all positive numbers total is +23.5

(+ 15 + 0.5 + 8) = + 23.5

Next subtraction values:

$$= (+15 + 0.5 + 8) - (-15 - 49)$$

Similar terms negative - value

= 23.5 − 64

(How do you solve all negative − value when they all need added up)

Added up all minus (-) symbol numbers.

$$\begin{array}{r} -15 \\ +-49 \\ \hline -64 \end{array}$$

There are two negative numbers (adding)

− 15 +(- 49) = - 64

Subtraction value:

$$\begin{array}{r} -64.0 \text{ (negative value)} \\ +23.5 \text{ (positive value)} \\ \hline -31.5 \text{ (subtract value −ve 31.5} \end{array}$$

Subtracted value = **23.5 − 64 = -31.5**

Order of Operations Methodology:

Many problems can't be solved in single step. Therefore, multistep problems required two or more steps before reach a solution. Multistep solution involves various arithmetic signs such as -, +, * , and /, square roots and so on.

Order of operations is very important step, which makes a great role when solve a multistep problem.

CBEST Math Section:
California Basic Educational Skills Test

The following order of operation is very important to remember.

First step 1:

First step is to solve inside numbers of parentheses **operation, which symbols ( ).**

Second step 2:

Second step is to solve powers value such as 2^2, 2^3, 2^x etc. ^{x} represents power value.

For example, $2^2 = 2 \times 2 = 4$

$2^3 = 2 \times 2 \times 2 = 8$

Third step 3:

Multiplication or/and division operations whatever comes first. Operation takes place from left to right in order.

Forth step #4

Finally, addition or/and subtraction operation, whichever comes first. Operation takes place from left to right is recommended.

CBEST Math Section:
California Basic Educational Skills Test

Multiplying operation:

Multiply

Multiplying two numbers result represents various words/phrases like product, times, twice, and so on.

Multiplying sign (x) or symbol * used in electronic calculator/computer.

x represents one of the multiply signs.

Multiply signs are a pair of parentheses ( ), *, letter 'x' etc.

An example of a pair of parentheses, 5(2+6), which can be multiplied this way;

$\{(5 \times 2) + (5 \times 6)\}$

$= (10) + (30)$

$= (10+30)$

$= 40$

Multiplying can be done both positive and negative number combined.

CBEST Math Section:
California Basic Educational Skills Test

Multiplying example of − and + values,

$$\begin{array}{r} -3 \longrightarrow -3 \\ +8 \longrightarrow *\,+8 \\ \hline -24 \end{array}$$ (negative 3 * positive 8)
(multiplication sign * or x)

- *+ results = -24
(Color is used here for readability purpose)

Let's figure out whether + or − sign's precedence of operation.

Either one, - Minus (-) or Plus (+) sign.

Look at the following:

Product of: (−) x (+) = - sign precedence

Product of: (-) x (-) = + sign -do-

Product of: (+) x (-) = - sign –do-

Product of: (-) x (-) x (-) = - sign and so on.

Product of: (-) x (+) x (-) = + sign

Learn from examples:

Example #1: Negative outcomes

Where x represents multiplication

(-3) x (+7) x (-5) x (-5) x (-2) x (-1)

Or -3 x +7 x -5 x -5 x -2 x -1

Or -1050 (use calculator)

Negative – 1050

CBEST Math Section:
California Basic Educational Skills Test

Example #2: Positive outcomes

Where * or x mark represents multiplication symbol or sign

(-3) x (+7) x (-5) x -5 x (-2) x (+1)

Or -3 x +7 x -5 x -5 x -2 x +1

Or 1050

1050 is not a negative value therefore it is Positive +1050

(No sign is necessary when outcomes positive values).

Division:

Division or divide represents number of ways:

Divided by, divided into, Ratio, half, quotient etc.

Dividing (/) Sign:

Dividing sign like this: / (slash) represents one of the dividing signs. Dividing sign operation, parentheses also use to group numbers i.e.

CBEST Math Section:
California Basic Educational Skills Test

$$5\left(\frac{2}{5}\right) = \frac{\cancel{5}}{1} \times \frac{2}{\cancel{5}} = \frac{2}{1} = 2 \text{ (reduce/lowest terms)}$$

So, final result is 2

Learn from Examples:

Dividing fractions

Example #1:

Divide the fraction, $\dfrac{-74}{-2}$

$$\frac{-74}{-2} = \frac{\cancel{-}37}{\cancel{-}1} = +37 = 37$$

(The operation involves two operation;

Lowest terms, and – divide – becomes +

value or (- / - becomes + precedence)

Example #2:

Divide the fraction, $\dfrac{-75}{25}$

$$\frac{-75}{25} = \frac{-3}{1} = -3$$

Simplify minus sign - and plus sign + issue.

CBEST Math Section:
California Basic Educational Skills Test

- divide + becomes − precedence

or

- / + becomes − precedence;

alternately,

+ / - becomes - precedence

Symbol Legend:

Minus sign represents - symbol

Add sign represents + symbol

CBEST Math Section:
California Basic Educational Skills Test

Measurement Ruler:

Introduction a ruler measurement sample, which often used by mathematic usage every day in our kids.

Understand Positive & Negative Numbers by using a Scale or Ruler.

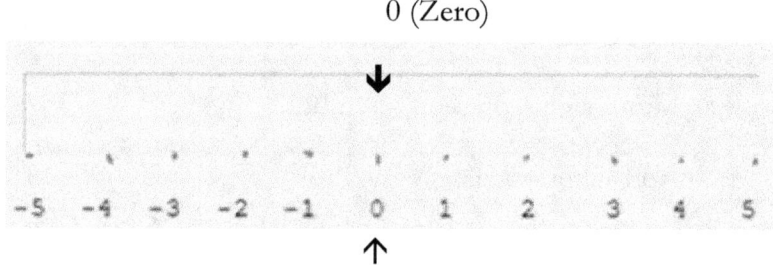

0 (Zero)

Arrow Pointing Zero

(Scale is not actual size)

Typical scale represents both negative (-) and positive (+) measurement, which 0 (zero) is located in the middle of the scale.

What is positive or negative number?

Negative number starts left side of the scale's 0 (zero) and positive number starts right side from the 0 (zero).

CBEST Math Section:
California Basic Educational Skills Test

Analyze Scale Elements:

The scale has two parts:

Left side negative starts 0 (zero) point through -5 shown above. This left side can be extended beyond -5 like -6, -7, -8 etc.

Similarly, right side of the scale starts from 0 (zero) to the right side and it continue up to 5 or +5. Both negative and positive scale sides can be extended up to infinity (symbol ∞).

This type of scale has many applications in the CBEST math test questions.

Math Properties:

Addition Axioms

1. Commutative

   Commutative does not make any difference, for instance, 4+6=6+4. It does not change total value.

CBEST Math Section:
California Basic Educational Skills Test

2. Associative

    Associative grouping such as (4+5), (6+3), (2+7) do not make any difference, for instance,

    (4 + 5) + 2 = 2 + (6 + 3)

3. Additive inverse

    Inverse means opposite number, for instance, 5 = 5 or 5 − 5 = 0, therefore, 5 and -5 are additive inverse.

4. Identity element

    Adding any number to 0 (zero) gives the original number. For example, 6+0=6,

    x+0=x

Multiplication Axioms

1. Commutative

    Commutative multiplication does not change value such as 4x6=6x4. However, it does not support for division.

CBEST Math Section:
California Basic Educational Skills Test

2. Association

    Associative multiplication grouping does not make any difference, for instance, (5x8)2=2(8x5). Both sides are balanced. However, division does not support associative multiplying.

3. **Multiplicative inverse**

    Multiplicative inverse is the reciprocal of the number. $4 \times \frac{1}{4} = 1$ (4 and $\frac{1}{4}$ action is inversely reduce lowest terms).

4. Identity element

    Identity element for multiplication is 1. For instance, 6=6x1, this does not change value. So, it gives original value.

Distributive Property

Distributive property processes distributing the number on the outside of the parentheses to each parenthesis to each number on the inside. For example, $z(x + y) = z(x) + z(y)$

Another example, $5(4+2) = 5(4) + 5(2)$

However, distributive property can't use to the following example,

$2(3. 4. 5) \neq 2(3). 2(4. 2(5)$

Where (.) dot sign represents multiplication operation, which is similarly meaning of parentheses operation of multiplication the example provide here above.

Sign $\neq$ represents not equal to (=)

$2(60) \neq 6.8.10$

Or $120 \neq 480$

The above operation does not support equation is balance; therefore, $\neq$ (not equal sign) has applied.

CBEST Math Section:
California Basic Educational Skills Test

Parenthesis operation should be done first (precedence). Look at the following operations:

$$120 \neq 480$$

120 becomes from the left side equation after solved this: 2(3. 4. 5)

$\neq$ Unequal sign

480 becomes from the right side equation after solved this: 2(3). 2(4. 2(5)

So, they are not balanced i.e. unequal.

CBEST Math Section:
California Basic Educational Skills Test

Fundamental Statistics:

Arithmetic **Mean** or **Average**

Average or mean is obtained by adding up few numbers, (5+6+8+9) = 28, and then divided that total number (28) by the total items of number, which is 4 items.

For example,

$$\text{Average} = \frac{(5+6+8+9)}{4} = \frac{28}{4} = 7$$

So, (5+6+8+9)/4 becomes mean value is 7

How do you find the average/mean number?

Step #1:

Add up all numbers, for instance, a group of numbers: 5, 6, 8, 9, and 12.

Step #2:

Divide by the number of items. As you've seen, the group contains 5 items.

Step #3:

Add up numbers (5+6+8+9+12) = 40

So, average calculation is,

$$\text{Mean} = \frac{\text{Adding Up All Group Numbers}}{\text{Group's Items Total}}$$

Group total = 40

Group's total items =5

Average/mean = (5+6+8+9+12) = $\frac{40}{4}$ = 10

Mean or average is 10.

Learn from Examples:

Example #1:

What is the average of 23, 27, 28, and 28?

A. 26.5

B. 27.6

C. 28

D. 29.4

E. 30

CBEST Math Section:
California Basic Educational Skills Test

$23+27+28+28 = 106$

106 becomes from adding up 4 items

Therefore, average $= \frac{106}{4} = 26.5$

The choice is A.

Example #2:

What is the mean of 0, 1, 10, 20, 21, 30, and 44?

A. 16

B. 17

C. 18

D. 19

E. 21

The group numbers are provided 0, 1, 10, 20, 21, 30, and 44.

Therefore, adding up them.

$0 + 1 + 10 + 20 + 21 + 30 + 44 = 126$

There are 7 items here above.

Therefore, mean $= \dfrac{\mathbf{126}}{\mathbf{7}} = 18$, The choice is C.

CBEST Math Section:
California Basic Educational Skills Test

CBEST PRACTICE TEST:

Question #1:

The average formula,

$$\text{Average} = \frac{\text{Adding Up All Group Numbers}}{\text{Group's Items Total}}$$

1. During a semester, a student received scores of 76, 80, 83, 71, 80, and 78 on six tests. What is the student's average score for these six tests?

    A. 76

    B. 77

    C. 78

    D. 79

    E. 80

$$\text{Average} = \frac{76+80+83+71+80+78}{6}$$

$$= \frac{468}{6} = 78$$

CBEST Math Section:
California Basic Educational Skills Test

Mode

Mode is the value that occurs most frequently in a given set of data. For example, mode value often occur statistical survey, for example, a set of group data can be test-scoring data record. The mode number is most frequently found in a group of numbers.

For example,

How do you find mode from data set provided below?

2, 9, 7, 9, 6, 4, 9, 3, 7, 9, 5, 7,3

With careful reading the listed data set, we have seen the following:

1. The number 9 appeared 4 times.

2. The number 7 appeared 3 times.

3. The number 3 appeared 2 times

4. The number 2 appeared 1 time

So, number 9 is appeared more often than other numbers provided. Therefore, the mode of this data set is 9.

Learn from Example:

Example #1:

A set of scores data: 3, 4, 5, 6, 6, 7, 8.

Find mode from the list below?

    A. 3

    B. 4

    C. 5

    D. 6

    E. 8

Mode is the data unit that appears most frequently. The umber 6 appeared twice therefore 6 is the mode of this data listing.

The number 6 appeared twice that's mode, the distribution of data provided is called **bimodal.**

Median

A median is simply the middle number of a group or a list of numbers after it has been organized in numbering ordered listing. In other word, a median literarily situated in the middle of a group or a list of numbers listing.

However, an exception is this, if a group of numbers contains an even number of items then averages the 2 middle numbers to get the median. It needs ability to read carefully written order listing of a set of data. For example, a set of scores data: 3, 4, 5, 6, 7, 8, and 9.

Which one is the median from the above listing data? The median number is 6, because it appeared middle of this data set; therefore, 6 is the median of this dataset.

CBEST Math Section:
California Basic Educational Skills Test

Learn from Examples:

Example #1: Even number of items' median.

Find the median of this set of data: 3, 4, 5, <u>6, 7</u>, 7, 8, 9.

These set of data statistic says two middle members i.e. <u>6 and 7.</u>

The given set of data contains two middle numbers are 6 and 7. Let's average them; 6+7=13/2=6.5 or $6\frac{1}{2}$.

The median number is $6\frac{1}{2}$.

Range

Rang is a set of different things or numbers of the same general type. In the arithmetic practice, arrange numbers in a row, for instance, a group of scores or numbers; subtract the smallest from the largest.

A group of scores: 4,3,2,7,9,13.

Find the range of the scores?

According to the definition, The range is (13 - 2) = 11

CBEST Math Section:
California Basic Educational Skills Test

Arithmetic Numerical Terms:

- Whole numbers
- Fractions
- Decimal
- Percent

Whole number

What are whole numbers?

Whole numbers are number that refers one or more digits which are 0 (zero) through 9 (nine). They (0 through 9) are arranged in a particular order. It means that whole numbers includes 0 (zero) and counting 1, 2,3,4,5,6,7,8,9,....999 and so on.

Place Value

What is place value?

A digit value is determined by its place, for example, 3907 is a four-digit number. This number contains four place values, for instance, it written convention is this: 3,907 (Three Thousand Nine Hundred Seven). It uses **a comma** after the digit 3, is (thousands place value) for readability purpose a comma (, a symbol) is used but does not need it. So, it is an optional separator.

[3,~~000~~] 3 is placed value is thousands;

[9~~00~~] 9 is placed value is hundreds;

[0~~0~~] 0 is placed value is tens;

[7] 7 is placed value is ones.

Learn from examples:

11 has <u>1</u> tens and <u>1</u> ones

25 has <u>2</u> tens and <u>5</u> ones

273 has <u>2</u> hundreds <u>7</u> tens <u>3</u> ones

9,054 has 9 thousands 0 hundreds 5 tens and 4 ones.

Large Numbers

534,850,729 in wording:

Five hundred Thirty-Four Million, Eight hundred Fifty Thousand, Seven hundred Twenty-Nine

CBEST Math Section:
California Basic Educational Skills Test

There are three breakdown steps to read it out.

5 3 4

    5 place value 100s

    3 place value 10s

    4 Place value 1s

8 5 0

    8 place value 100s

    5 place value 10s

    0 place value 1s

7 2 9

    7 place value **100**s

    2 place value **10**s

    9 place value **1**s

CBEST Math Section:
California Basic Educational Skills Test

Rounding Off Concept:

How do you rounding off any number?

It depends on nearest million, thousand, hundred, and ten.

Round to nearest thousand examples:

Nearest thousand rounding off 545,685 number becomes 546,000.

You asked why?

The number is 545,685

In words it is written this,

"Five hundred Forty five thousand, Six hundred Eighty Five"

It has two parts,

First part 545,~~685~~

Second part ~~545,~~685

The underlined part is 685 that is fall into 1000. Because, after first part 500 or above 500 considered round off figure i.e. 1000 (thousand). Therefore, 685-number becomes 1000, which then added to the previous digit(s) 545,xxx that becomes 546,000.

CBEST Math Section:
California Basic Educational Skills Test

ROUNDING OFF METHODOLOGY:

Step #1:

Locate the 'Place Value' in which you want to rounding-off.

[The above example shown (looking round-off place value thousand), the 'Place Value' ~~545,~~685 to be rounding off]

Step #2:

Look at the immediate right, one digit after comma, place value. Recall above example, ~~545,~~685. Its place value is underlined number (6xx).i.e. 6

Step #3.

From the step 2 analysis, we have identified 'Place Value' is 6.

Step #4.

Rule of thumb of "rounding-off" is that if the place value identified **5** or higher then round the 'Place Value' up **1**.

Step #5.

If the identified number **less** than **5** or 4 or less than 4 then leaves the left side value $(545,\mathbf{X}\text{xx})$ unchanged and make all the other numbers to its right to zeroes.

Xxx, where the big X represents less than 5.

Detail Descriptions of 'Place Value'

Billion numbers

1,000,000,000	Billion
1,000,000	Million
1,000	Thousand
100	Hundred
10	Ten
1	One
.	Decimal point Symbol is dot
.10	Tenths
.01	Hundredths
.001	Thousandths
.0001	Ten-thousandths
.00001	Hundred-thousandths
.000001	Millionths
.0000001	Ten-millionths
.00000001	Hundred-millionths

Cost Estimating Method:

A cost estimate is the approximation value of a product or something valuable tangible thing, for instance, cost estimate is a product price (value) that may uses (rounding-off figure) rounded numbers to estimate the following cases; sums (+), differences or minus (-), multiplication (x), and quotient (/).

Estimating Sums (+):

Rounded numbers use to estimate sums (+)

thousand:

For example,

6,143 ≈ 6,000 (nearest thousand)

2,421 ≈ 2,000 (nearest thousand)

Estimate value ---> 8,000 (rounded value)

CBEST Math Section:
California Basic Educational Skills Test

Symbol ≈ represents approximately equal to

Rounded numbers use to estimate sums (+) thousand:

For example,

$$5,143 \approx 5,000 \text{ (nearest thousand)}$$
$$\underline{2,721 \approx 3,000 \text{ (nearest thousand)}}$$

Estimate value ---> 8,000 (rounded value)

Symbol ≈ represents approximately equal to

Estimating Differences (-):

Estimating difference value rounded number hundred:

$$\text{Actual}\downarrow \text{-----------}\downarrow \text{Estimated}$$
$$810,554 \longrightarrow 800,000$$
$$\underline{\text{Subtracting } 325,611 \longrightarrow -300,000}$$

Estimated differences value 500,000

CBEST Math Section:
California Basic Educational Skills Test

Estimating difference value rounded number **hundred**:

Actual↓ -------- ↓Estimated value

729,542 --> 700,000

Subtracting 216,611 --> - 200,000

Estimated differences value 500,000

---

hundred:

479 estimated value 500

925 estimated value 900

Estimated product of (500x900) 450,000

So, we estimated (479x925) ≈ 450,000

Calculate estimating rounded number **hundred**:

579 estimated value 600

825 estimated value 800

Estimated product of (600x800) 480,000

CBEST Math Section:
California Basic Educational Skills Test

Examples:

Estimating Products/Multiplications Place Value Nearest Hundred

755 --> 700 (Rounded one number down)

450 --> 500 (Rounded one number up)

350,000 (estimating products)

So, 755 x 450 ≈ 350,000 (Estimating Rounding off)

```
Value Nearest Hundred

755 --> 800 (Rounded one number up)

450 --> 400 (Rounded one number down)

    320,000 (estimating products)

So, 755 x 450 ≈ 320,000    (Estimating
Rounding off)
```

Rounding off above examples, one number down and up two instances produced two results i.e. 350,000 and

CBEST Math Section:
California Basic Educational Skills Test

320,000. They are very closer approximation than rounding both numbers up.

Therefore, it makes sense of the above two demonstration up and down and vice versa that rule standard rule. This standard rule math is frequently seen at real CBEST Math test. It is recommended to study it very closely.

Estimating Quotients (divide):

Use rounded numbers to estimate quotients by rounding nearest <u>hundred</u>.

For example nearest hundred,

581 --> 600 (Nearest hundred rounding off)

<u>218 --> 200 (Nearest hundred rounding off)</u>

3 (Estimated quotients)

So, $581/187 \approx 3$

```
For example nearest hundred,

581 --> 600 (Nearest hundred rounding off)

187 --> 200 (Nearest hundred rounding off)

            3 (Estimated quotients)
So, 581/187 ≈ 3
```

Fraction

Fraction is a numerical quantity that is not a whole number e.g. 0.5, $\frac{1}{2}$, $\frac{3}{4}$ etc.

Formula of fraction:

$$\text{Fraction} = \frac{\text{Numerator}}{\text{Denominator}} ; \text{Fraction} = \frac{1}{4}$$

The dividing line in the middle of a fraction that represents division thus it means 1 out of 4. Therefore, it also represents "Out Of"

Our fraction example,

One quarter or $\frac{1}{4}$, this fraction has two parts; 1 and 4. The top number is 1, which is known as **numerator** and the lower part number is 4, which is known as **denominator.**

CBEST Math Section:
California Basic Educational Skills Test

By definition, a fraction whose **denominator** is a power of **ten** and whose numerator is expressed by figures placed to the right of a decimal point.

In practical application, fraction can be illustrated all most all objects such as a circle, rectangle area, squared box/area, an apple cut, and so on.

So, fraction $= \frac{1}{4} = .25 = 25\%$ is not a whole number.

Typical example of $\frac{1}{4}$ pie fraction

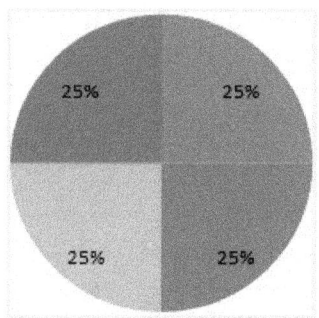

Circle 1/4 Fraction

The circle creates $4 \times \frac{1}{4} = 1$ or 100% whole circle.

Multiplication operation is denoted x.

CBEST Math Section:
California Basic Educational Skills Test

Examples of Fraction:

Whole giant squared area shown below is an example how looks like a square. Its four sides are equally same length.

Square fraction ¼ each.

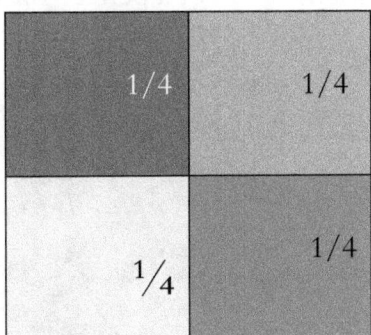

Whole Rectangle

CBEST Math Section:
California Basic Educational Skills Test

Rectangle Fraction ½ each

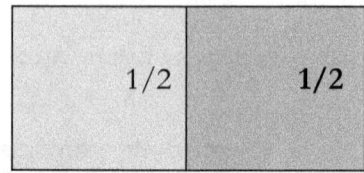

SQUARES NUMBER:

What is square number?

Square is a number that multiplies itself, for example, 4 x 4 or 4^2 = 16 squared box.

An illustration of a perfect 16 sqft area

16 squared is a perfect squared area, which is a whole number. It is called 4^2, which is an exponential number.

CBEST Math Section:
California Basic Educational Skills Test

Exponential Number:

Exponent numbers can multiply by itself many times. Few examples of exponential numbers are; 4^2, 4^3 8^2 8^3 and their values are analyzed herein:

$4^2 = 4 \times 4 = 16$

$4^3 = 4 \times 4 \times 4 = 64$

$8^2 = 8 \times 8 = 64$

$8^3 = 8 \times 8 \times 8 = 512$

A list of Perfect squared

$1^2 = 1$	$11^2 = 121$
$2^2 = 4$	$12^2 = 144$
$3^2 = 9$	$13^2 = 169$
$4^2 = 16$	$14^2 = 196$
$5^2 = 25$	$15^2 = 225$
$6^2 = 36$	$16^2 = 256$
$7^2 = 49$	$17^2 = 289$
$8^2 = 64$	$18^2 = 324$
$9^2 = 81$	$19^2 = 361$
$10^2 = 100$	$20^2 = 400$

Classification of Fraction:

Analysis of Fraction:

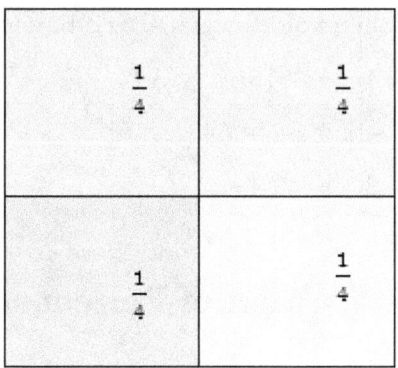

Common Fraction:

There are four pieces of $\frac{1}{4}$ (one-fourth) makes a giant square box.

Fraction $=\dfrac{1}{4}$ in this example, the upper number is 1, and the denominator is 4. It means 1 represents numerator and 4 represents denominator.

It is a fraction value of 1/4, which makes it a *common fraction*. Common fraction's denominator value is always greater than the numerator such as $^1/_4$, $^3/_4$, 5/6 and so on.

Improper Fraction:

Second type of fraction is *improper fraction*. What is improper fraction? Well, when numerator is larger than the denominator such as 4/3, 3/2, 6/5 and so on.

Example analysis of improper fraction:

Improper fraction $=\frac{26}{7}$, which can be converted into a mixed fraction $= 3\frac{5}{7}$. Look at the following paragraphs for more information about "Mixed Numbers".

Mixed Numbers:

What is mixed number?

Improper number like this $3\frac{5}{7}$ is an example of mixed number, which was discussed earlier.

A mixed bombers example is this: $3\frac{5}{7}$.

In this example, there are two parts; first part is 3, which is known as a whole number and remaining number contains $\frac{5}{7}$ a numerator 5 and a denominator 7. So, when a term contains both a fractional number like this $\frac{5}{7}$ and a whole number like 3, which created a mixed numbers ($3\frac{5}{7}$).

Learn from examples:

Recall previously used improper fraction numbers, $\frac{26}{7} = 3\frac{5}{7}$

Question is how do you change an improper fraction numbers into a mixed numbers?

A mixed fraction is $3\frac{5}{7}$.

Let change it an improper fraction. There is a formula that you can use. It is provided herein.

The whole number multiplying (x) denominator adds up (+) numerator.

So, converting mixed number $3\frac{5}{7}$ into improper fraction is simply convention like this:

As per formula, 3 (whole number) x 7 (denominator) + 5 (numerator) = 26, which becomes numerator. And the denominator is remained 7. So, it is now written expression is $\frac{26}{7}$.

Mixed numbers $3\frac{5}{7}$ becomes improper numbers $\frac{26}{7}$

Reduce Fractions Lowest Terms:

Reducing a fraction numbers to its lowest terms is simply dividing the numerator and denominator by the largest number that divides evenly into both (numerator & denominator).

For example,

Reduce $\frac{20}{24}$ by dividing both terms by 4 to get $\frac{5}{6}$

How's it? $\frac{20}{24}$ Divide $\frac{4}{4} = \frac{5}{6}$

$\frac{20}{24}$ Reduced lowest fraction is $\frac{5}{6}$

Another example,

Reduce $\frac{12}{14}$ divide $\frac{2}{2} = \frac{6}{7}$ (dividing numerator and denominator by 2.

$\frac{12}{14}$ Reduced lowest fraction is $\frac{6}{7}$

Add Fractions:

Adding up fractions involves two more common or/and improper fraction numbers. First, change all denominator values into theirs lowest common denominator (LCD). How do you get LCD?

LCD or lowest common denominator number can be obtained by divided evenly by all the denominators (all fractional numbers' denominator involved).

For example

Adding up two fractions, $\frac{5}{6}$ plus $\frac{3}{8}$.

$$\frac{5}{6} + \frac{3}{8} = \{(\frac{4}{4} \times \frac{5}{6}) + (\frac{3}{3} \times \frac{3}{8})\} = \{\frac{10}{24} + \frac{12}{24}\}$$

$$or \quad \frac{20+9}{24} = \frac{29}{24} = 1\frac{5}{12}$$

CBEST Math Section:
California Basic Educational Skills Test

Let analysis of the previous equation:

How's work?

Change their lowest common denominators

Change ($\frac{5}{6}$ X $\frac{4}{4}$) equal to = $\frac{20}{24}$;

And change ($\frac{3}{8}$ X $\frac{3}{3}$) equal to = $\frac{9}{24}$.

Now both fractions have the same LCD 24.

So, $\frac{29}{24}$, it's an improper fraction.

How do you change improper fraction into a mixed fraction?

Dividing numerator 29 by denominator 29, the dividing result becomes 1 $\frac{5}{12}$ as a mixed numbers.

Example #2:

Problem Adding up = $\frac{1}{3}$ + $\frac{1}{4}$

Adding, $\frac{1}{3}$ + $\frac{1}{4}$ = ($\frac{1}{3}$ x $\frac{4}{4}$) + ($\frac{1}{4}$ x $\frac{3}{3}$)

Make both fractions' denominator same by multiplying respectively 4 and 3, which becomes 12 LCD.

CBEST Math Section:
California Basic Educational Skills Test

Or $\dfrac{4}{12} + \dfrac{3}{12}$

Or $\dfrac{(4+3)}{12} = \dfrac{7}{12}$

So, both fractions become $\dfrac{7}{12}$ a common fractional number.

Adding Mixed Numbers:

Adding up, $3\dfrac{1}{3} + 5\dfrac{2}{3}$, both are mixed numbers.

Adding mixed numbers method is four steps.

First step,

Figure out the LCD. It's 3 from the above equations provided.

Second step,

Adding up the two whole numbers, (3+5)=8.

Third step,

Adding up fractions parts together; $\dfrac{1}{3} + \dfrac{2}{3} = \dfrac{1+2}{3} = \dfrac{1}{1} = 1$

Fourth Step,

Finally, adding up both (whole numbers) 8 + (fractions total) 1 = 9. .

Learn from Examples:

$5\frac{1}{3} + 4\frac{2}{3}$.

These two mixed numbers have two common fractional parts. They are 1/3 and 2/3, which can be written

$(\frac{1}{3} + \frac{2}{3})$

Or $(\frac{2+1}{3}) = \frac{3}{3} = 1$

Next, there are two whole numbers (5+4) = 9.

Two fractions total =1

So, adding up, $5\frac{1}{3} + 4\frac{2}{3}$ becomes = 8+1 = 9.

Subtraction Mixed Numbers:

Subtraction mixed numbers method is 3 steps:

First,

Find the LCD of all common fractional numbers portion and make its total.

Second, subtract fractional part that involves subtraction.

Third, if fraction requires borrowing from whole numbers (subtract from whole numbers when fraction produce negative (-) fraction) then subtract borrowing portion from the whole numbers. For instance, $(1\frac{1}{3} - \frac{2}{3}) = 1 - \frac{1}{3} = \frac{2}{3}$

So, $(1\frac{1}{3} - \frac{2}{3})$ becomes $\frac{2}{3}$

Learn from Examples:

Example #1:

Subtraction, $5\frac{2}{3} - 3\frac{1}{3}$

This equation has 3 steps:

Step 1, subtraction (5-3) = 2

$5\frac{2}{3} - 3\frac{1}{3}$

Step 2, Figure out their LCD. It's 3.

$= (\frac{2}{3} - \frac{1}{3}) = \frac{2-1}{3} = \frac{1}{3}$

Step 3,

CBEST Math Section:
California Basic Educational Skills Test

Adding up, whole number from step 1 and step 2 together, $2\frac{1}{3}$.

Note that subtracting from $\frac{2}{3}$ to $-\frac{1}{3}$, which outcomes positive fraction of $\frac{1}{3}$. Because $\frac{2}{3}$ is greater than from $\frac{1}{3}$; therefore, just place this fraction in front of subtracted whole number 2 (recalled step 1 outcomes).

So, $5\frac{2}{3} - 3\frac{1}{3}$ becomes a mixed number $2\frac{1}{3}$.

Example #2:

Substation, $5\frac{1}{3} - 3\frac{2}{3}$

This subtraction is little different than previous example because two fractional parts were just opposite.

$(\frac{1}{3} - \frac{2}{3}) = -\frac{1}{3}$, which is a negative value.

However, $5\frac{1}{3}$, which has whole number is 5 that is greater value than -3. This -3 comes from $-3\frac{2}{3}$, which to be subtracted from positive fraction of $5\frac{1}{3}$.

89

CBEST Math Section:
California Basic Educational Skills Test

The equation has positive whole number (5-3) =2; however, fraction subtraction result is $-\frac{1}{3}$.

Therefore $-\frac{1}{3}$ to be subtracted from the 2 (whole number).

Thus, $2 - \frac{1}{3} = \frac{6-1}{3} = -\frac{5}{3} = 1\frac{2}{3}$

So, $5\frac{1}{3} - 3\frac{2}{3}$ becomes $1\frac{2}{3}$

Alternate Method:

$5\frac{1}{3} - 3\frac{2}{3}$ (figure out LCD i.e. 3

$= \frac{16}{3} - \frac{11}{3}$ (subtracting numerator results)

$= \frac{(16-11)}{3} = \frac{5}{3}$ (Improper fraction)

$= 1\frac{2}{3}$ (change to mixed numbers).

Subtracting mixed numbers:

Alternately equation is subtracted $(\frac{1}{3} - \frac{2}{3}) = \frac{1-2}{3} =$

$-\frac{1}{3}$ (negative 1/3, which needs subtraction from whole number)

Subtraction whole numbers: (5-3)= 2

CBEST Math Section:
California Basic Educational Skills Test

Now, it needs subtract negative fractional portion i.e. $-\frac{1}{3}$ (negative value) has to be subtracted from the whole number.

Therefore, $2 - \frac{1}{3} = (\frac{2}{1} - \frac{1}{3}) = \frac{(6-1)}{3} = \frac{5}{3} = 1\frac{2}{3}$

So, subtracting result is $1\frac{2}{3}$

Subtracting Fractions:

All fractions addition and subtraction involved LCD.

Common fraction:

Example #1:

5/8 same as $\frac{5}{8}$ similarly 1/8 same as $\frac{1}{8}$

Subtracting $5/8 - 1/8 = 4/8 = \frac{1}{2}$

Example #2:

First find LCD,

$3/4 - 2/3$

(Lowest common denominator is (4 * 3) = 12

CBEST Math Section:
California Basic Educational Skills Test

$$= \frac{\{(3x3)+(4x2)\}}{12}$$

$$= \frac{(9+8)}{12}$$

$$= \frac{17}{12}$$

$$= 1\frac{5}{12}$$

Multiplying Fractions:

Method of multiplication of fraction is simple.

Step #1:

Reduce to all fractions to lowest terms.

Reduce terms, $\frac{3}{7} \times \frac{5}{18}$ becomes $\frac{1}{7} \times \frac{5}{6}$

Step #2:

Cancel multiply fractions.

Lowest cancellation product that becomes $\frac{1}{7} \times \frac{5}{6}$

Step #3:

After cancel multiplying fractions, multiply horizontally all reaming numbers.

CBEST Math Section:
California Basic Educational Skills Test

Multiplying remaining numbers is, $\frac{5}{42}$

The final multiplying product is a common fraction $\frac{5}{42}$

Learn from Examples:

Solve the product of $\frac{2}{3} \times \frac{1}{4}$

$$\frac{2}{3} \times \frac{1}{4} = \left(\frac{\cancel{2}}{3} \times \frac{1}{\cancel{2} \times 2}\right) = \frac{1 \times 1}{3 \times 2} = \frac{1}{6}$$

So, $\left(\frac{2}{3} \times \frac{1}{4}\right)$ reduced to $= \frac{1}{6}$

(x represents multiplying symbol)

Rule of thumb of multiplying fractions must be reduced to its lowest terms possible. So, we reduced it to $\frac{1}{6}$. It then can't reduce further lowest terms thus multiplying fraction is ended here.

    The above instance is, cancelling multiplying factors inversely. It illustrates divide evenly into one numerator (2) and denominator (2) and then reaming numbers multiplying horizontally (multiplying two numerator into one number and similarly multiplying two denominators into one number); $\frac{1 \times 1}{3 \times 2} = \frac{1}{6}$

CBEST Math Section:
California Basic Educational Skills Test

Mixed Numbers Multiplying method:

Multiplying mixed numbers requires several steps, for example, first change all the mixed numbers terms into improper fractions that explain hereafter step by step.

Step #1:

Multiply the whole number by the denominator of the function. For instance, a mixed number is $2\frac{1}{2}$; 2 x 2 = 4

Step #2:

Add up step #1 product with numerator of the fraction.

For instance, 4+1 = 5

Step #3

Number 5 (from step #2) is now become numerator of the fraction.

CBEST Math Section:
California Basic Educational Skills Test

Step #4:

The denominator has no change i.e. 2.

Therefore, mixed number, $2\frac{1}{2}$ becomes $\frac{5}{2}$

So, improper number is $\frac{5}{2}$

Mixed Numbers Dividing methodology:

Divide fraction involves two fractions that one divides another. Divide common fraction by another common fraction is very straight forward process.

For example, $\frac{1}{7}$ divide by $\frac{1}{7} = (\frac{1}{7} \times \frac{7}{1}) = \frac{7}{7} = 1$

The dividing process, it inverts the second fraction turn upside down like this $\frac{1}{7} \times \frac{7}{1}$ and thereafter simplify the terms, which resulted 1.

What's about mixed numbers dividing by another mixed numbers?

The process is same as above example, but the exception, first convert mixed fraction into improper fraction.

CBEST Math Section:
California Basic Educational Skills Test

For example,

Step #1:

$2\frac{1}{2}$ divide $1\frac{1}{2}$ (Turn mixed terms into improper fraction separately).

Step #2:

$\frac{5}{2}$ Divide $\frac{3}{2}$ (they are now improper fraction)

$\frac{5}{2} \times \frac{2}{3} = \frac{5}{3}$ (Improper fraction);

$\frac{5}{3} = 1\frac{2}{3}$ (improper fraction converted mixed numbers).

Fraction Decimal:

What is fraction decimal?

Fraction decimal is written by using a decimal point. Fraction numbers are located to the right of the decimal point are fractions with 0.1 or $\frac{1}{10}$; 0.01 or $\frac{1}{100}$; 0.001 or $\frac{1}{1000}$ and so on. It means points of decimal are fractions with denominators of 10, 100, 1000, and 1000.

CBEST Math Section:
California Basic Educational Skills Test

Comperision: Decimals and fractions

Decimal --------- Fraction-------Pronunciation

Decimal point

↓ ↓ ↓

•

Decimal	Fraction	Pronunciation
.9	$\frac{9}{10}$	Tenths (Nine tenths)
.09	$\frac{9}{100}$	Hundredths
.009	$\frac{9}{1,000}$	Thousandths
.0009	$\frac{9}{10,000}$	Ten-thousandths
.00009	$\frac{9}{100,000}$	Hundred-thousandths
.000009	$\frac{9}{1,000,000}$	Millionths
.0000009	$\frac{9}{10,000,000}$	Ten-millionths
.00000009	$\frac{9}{100,000,000}$	Hundred-millionths

CBEST Math Section:
California Basic Educational Skills Test

Decimal ↓	Fraction ↓	Lowest Terms ↓
.6	$\frac{6}{10}$	$\frac{3}{5}$
.7	$\frac{7}{10}$	N/A
.8	$\frac{8}{10}$	$\frac{4}{5}$
.9	$\frac{9}{10}$	N/A
.09	$\frac{9}{100}$	N/A
.009	$\frac{9}{1000}$	N/A
.0009	$\frac{9}{10000}$	N/A
.00009	$\frac{9}{10000}$	N/A

CBEST Math Section:
California Basic Educational Skills Test

Decimal ↓	Fraction ↓	Lowest Terms ↓
.1	$\frac{1}{10}$	N/A
.2	$\frac{2}{10}$	$\frac{1}{5}$
.3	$\frac{3}{10}$	N/A
.4	$\frac{4}{10}$	$\frac{2}{5}$
.5	$\frac{5}{10}$	$\frac{1}{2}$
.6	$\frac{6}{10}$	$\frac{3}{5}$
.7	$\frac{7}{10}$	N/A
.8	$\frac{8}{10}$	$\frac{4}{5}$
.9	$\frac{9}{10}$	N/A
.09	$\frac{9}{100}$	N/A
.009	$\frac{9}{1000}$	N/A
.0009	$\frac{9}{10000}$	N/A
.00009	$\frac{9}{10000}$	N/A

CBEST Math Section:
California Basic Educational Skills Test

Decimal	Fraction	Pronunciation
.		Decimal point
.1	$\frac{1}{10}$	Tenths
.01	$\frac{1}{100}$	Hundredths
.001	$\frac{1}{1,000}$	Thousandths
.0001	$\frac{1}{10,000}$	Ten-thousandths
.00001	$\frac{1}{100,000}$	Hundred-thousandths
.000001	$\frac{1}{1,000,000}$	Millionths
.0000001	$\frac{1}{10,000,000}$	Ten-millionths
.00000001	$\frac{1}{100,000,000}$	Hundred-millionths

Decimal

What is decimal?

The decimal describes the base-10 number system. Base-10 refers to the numbering system in widely used. For example, a number 547, it describes; base-10 refers respectively 5 value place is hundred's place, 4 value place is ten's place, and 7 value place is one's place. It also describes that each number is 10 times the value to the right of it, hence the term base-10. The decimal number 0.29 describes 0.20 + 0.09, which each number is 10-time the value to the right. It means 2 is tenth value place i.e. 0.2 or $\frac{2}{10}$ or $\frac{1}{5}$ fraction. 1 is known as **numerator (divisor)** and 5 known as **denominator (divider)**.

And 9 is hundredth (0.09) value place i.e. $\frac{9}{100}$ or $\frac{9}{100}$ fraction. 9 known as **numerator** and 100 is known as **denominator**.

CBEST Math Section:
California Basic Educational Skills Test

Decimal refers a system of numbers and arithmetic based on the number ten, tenth, and power of ten.

For example,

1 hundred = 10 tens.

1 ten = 10 ones

1 one = 10 tenths

1 tenth = 10 hundredths

The above examples are units that divided into ten equal parts, which known as '**decimal system**'.

Number, currency (dollar), and metric measurement are decimal system, which all are decimal system i.e. each unit is divided into ten equal parts.

For example,

Decimal Currencies:

1 dollar = 10 dimes = 100 pennies

$5.29

$ dollar symbol

CBEST Math Section:
California Basic Educational Skills Test

How do you analyze $5.29?

$5.29 = 5 whole number (dollar) next to it is a decimal point (.) that followed by 2 dimes (2 tenth of a dollar) that followed by 7 pennies/cents (7 hundredths of a dollar).

U.S. currency coins are widely used as following convention system:

Writing money convention, it writes cents or pennies as the two decimal places. Two decimal places are tenth or 1/10 or 0.01. A leading 0.~~00~~ (zero) is written to the left, it puts before the decimal point if there is not whole number. See following currency examples of U.S. currencies.

One penny/cent is denoted: $0.01
One nickel is denoted: $0.05
One dime is denoted: $0.10
One quarter is denoted: $0.25
Half dollar is: $0.50
One dollar is: $1.00

Note:
Currency does not use more than two digits decimal point. For example, one dollar ten cents is written as $1.10. However, currency calculation usually rounded off round figure.

Decimal Point Measurement:

Decimal uses in Length, width, height, for example, tenth, hundredths, and thousandths.

Leading point (.) starts numbers.

The First three decimal places

```
         1 tenth              1 hundredth           1 thousandth
  . _____         _____      _____

0.1 (1 tenth)    0.01 (1 hundredth)    0.001 (1 thousandth)
```

An Example of 1 tenth shown below

1 + 1+ 1 +1 +1+1+1+1+1 +1 =10

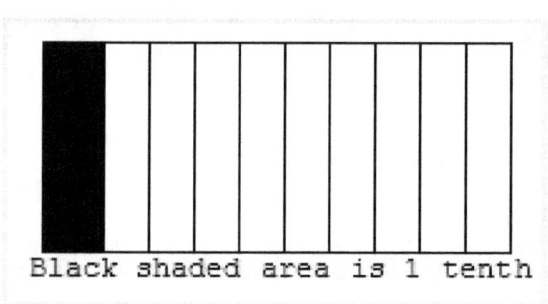

Black shaded area is 1 tenth

The rectangle is equally divided 10 parts.

Each part is = 1/10 = 0.1 part

CBEST Math Section:
California Basic Educational Skills Test

The black shaded portion represents 1 part of the rectangle, which is 0.1

An example of 1 hundredth (0.01)

Red box bellow is denoted 0.01.

It means 1/100

	2	3	4	5	6	7	8	9	10
2									
3									
4									
5									
6									
7									
8									
9									
10									

Each side 10 box x 10 box = 100 box

Colored box is 1/100 = 0.01 or 1 hundredth

Cube

Example of 1 thousandth = 1/1000

or 0.001

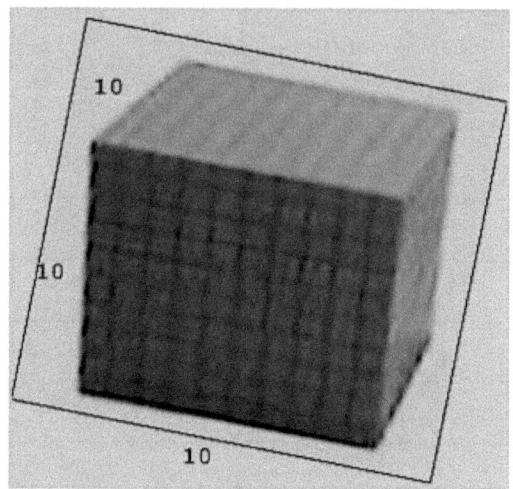

Assume each side consists of 10 small cubes.

10 x 10 x 10 = 1000

So, one small cube is 1/1000 = 0.001

or 1 thousandths

CBEST Math Section:
California Basic Educational Skills Test

Examples:

1. 0.5 or 5 tenths (5 is tenth place value)

2. 0.50 (5 is still tenth place but zero has no value)

3. 0.05 (5 moves to hundredth place value)

4. 0.09 or 9 tenths (9 is hundredths place value)

5. 0.05 or 5 hundredth (5 is hundredth place value)

6. 0.005 Or 5 thousandths (5 is thousandth place value)

7. 0.0001 is 1 ten-thousandths

8. 0.00001 is 1 hundred-thousandths

9. 0.000001 is 1 millionths [{1 is 6th placed (6-position after decimal point) value after leading zero i.e. (0.xxxxx1)}]

    Note: Measurement calculation is practically use up to thousandths (0.xxx) place value.

    x represents any valid numerical digit.

CBEST Math Section:
California Basic Educational Skills Test

Whole Number:

Rounding Nearest One to a <u>Whole Number</u>:

Examples:

The symbol ~ means approximately equal to (=)

1. 5.5 ~ 6
2. 17.26 ~ 17
3. 15.537 ~ 16

Rounding <u>Nearest Tenth</u> Place:

Examples

1. 0.25 ~ 0.3
2. 6.63 ~ 6.6
3. 9.381 ~ 9.3

Rounding <u>Nearest Hundredths</u> Place:

Examples

1. 0.476 ~ 0.48
2. 2.383 ~ 2.38
3. 4.246 ~ 4.25

CBEST Math Section:
California Basic Educational Skills Test

ADDING & SUBTRACTING DECIMALS

Adding decimals:

To add two or more decimals, it needs align (line up) the decimal points and then add in the same manner.

Example #1: Adding decimal

```
Look at red spot ↓

For example, 19.51

               26.26
               _____
               45.77
```

Example #2: Adding decimal point

Look at decimal point marked it ↓ red spot

$$19.51$$
$$+\ 22.67$$

Total   **42.18**

Subtracting Decimals

To subtract two decimals, it needs align (line up) the decimal points and then subs tract them.

Example #1: Subtraction

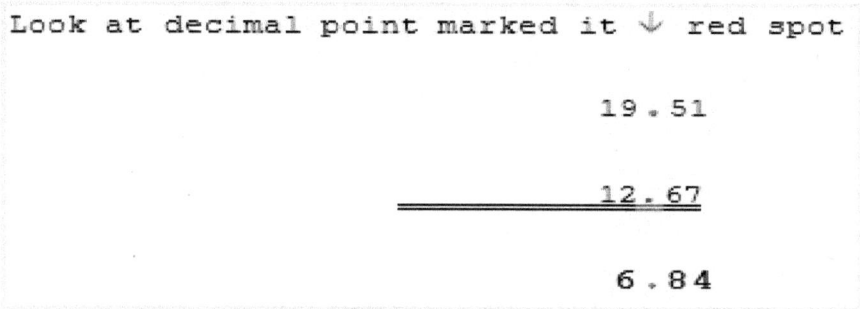

```
Look at decimal point marked it ↓ red spot

                              19.51
                           __ 12.67
                               6.84
```

Example #2: Subtraction

Look at decimal point marked it ↓ red spot

$$79.51$$

$$\underline{27.17}$$

52.34

CBEST Math Section:
California Basic Educational Skills Test

Multiplying Decimals

How to multiplying decimals?

It is pretty easy; first, count all number of digits that located after the decimal points.

For instance,

```
Aligning ↓ decimal points

    .01  (two digits after the decimal point)

    .15  (two digit after the decimal point)
   ─────────────────────────────────────────

    .0015 (multiplying results)
```

First, multiply two numbers.

Second, write the product, which comes after multiplying.

Third, count the numbers (like this .01 is two digits + .15 is also two digits), which counts total 4 digits. Remember, "Place Value" after decimal points that begin like this convention; .1 (tenth), .01 (hundredth), .001 thousandths and so on. So, move product (multiplying results) to the right as many

111

places as necessary i.e., it based on how many numbers count after decimal points of the two numbers that about to produce its product.

Fourth, place the decimal point before write the product results, which come from multiplying outcomes.

Example #1:

Aligning ↓ decimal points

0.01 (two digits after the decimal point)

<u>0.015 (two digit after the decimal point)</u>

0.00015 (multiplying results)

CBEST Math Section:
California Basic Educational Skills Test

Dividing Decimals:

Dividing decimals number involves divisor (numerator) and divider (denominator).

For example,

When, Divisor = 5.00

Divider = 1.25

It looks like this $\dfrac{5.00}{1.25} = \dfrac{500}{125} = 4$

Alternately,

$$1.25 \overline{)5.00} = 125 \overline{)500}^{\,4}$$

Learn from Example:

Example #1:

(Multiply 100 both divisor and divider)

Divisor = 3.03 = 303

Divider = .02 = 2

First, both numbers contain two digits decimals. Thus, if we multiply both of them by 100 then the value will not change. Because, both place values are .03 and .02 hundredths Place Value.

So, 3.03 becomes 303 and .02 becomes 2.

$$\frac{303}{2} = 151.5 \text{ or } 151.50$$

(An extra zero like this .50 will not change any value; however, .05 has great impact of value change).

Percent/Percentage (%)

What is percent?

In math calculation, the symbol % sign represents percentage.

Percent refers to the number of parts out of 100 equal parts.

In other word:

Percent means parts per 100 equal parts.

For example, if the number is one part out of 100 equal parts then it is known as 1 percent or 1%. percent usages symbol % followed by a number.

CBEST Math Section:
California Basic Educational Skills Test

Learn Percent from Examples:

What is 100%?

100% stands for a whole object.

For example, a circle given bellow is a complete circle object.

The circle is divided into 3 parts in which 33 parts is blue, 30 parts green, and 37 parts brown color. Thus, this (circle) represents 33%, 33%, and 37%.

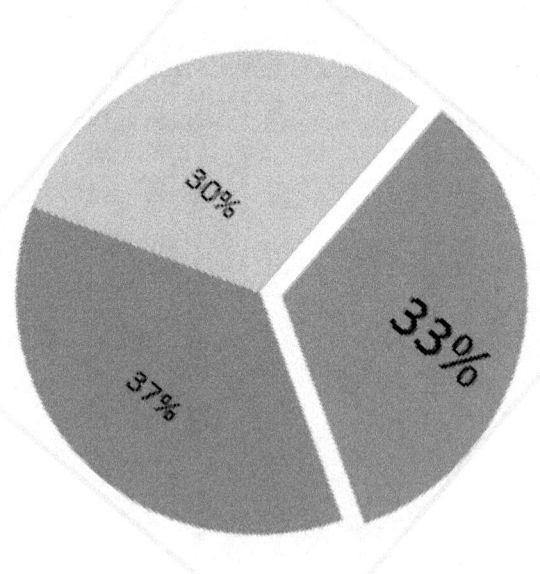

Typical Pie graphs along with percentage sign.

Again, the whole circle object is 100%, which made of 30%+37%+33% respectively green, brown, and blue colored pie.

Changing Percent to Decimal Point:

To change percent to decimal, it needs just dividing by 100.

For example,

$$33\% = \frac{33}{100} = .33 \text{ (33\% becomes .33)}.$$

So, .33 is the equivalent of 33 part out of 100 or $\frac{33}{100}$

Examples illustrated percent, decimals, and fractions that representing the following diagram.

CBEST Math Section:
California Basic Educational Skills Test

Examples illustrated percent, decimals, and fractions that representing the following diagram. |

$30\% = .30 = \frac{30}{100}$

$33\% = .33 = \frac{33}{100}$

$37\% = .37 = \frac{37}{100}$

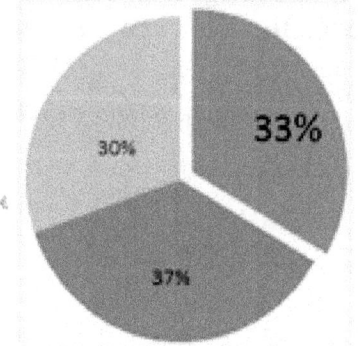

$100\% = 100 =$ Whole is 1.

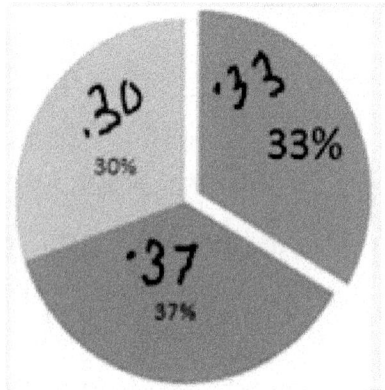

$.30 = 30\% = \frac{30}{100}$

$.33 = .33\% = \frac{33}{100}$

$.37 = .37\% = \frac{37}{100}$

$100 = 100\% =$ Whole is 1.

CBEST Math Section:
California Basic Educational Skills Test

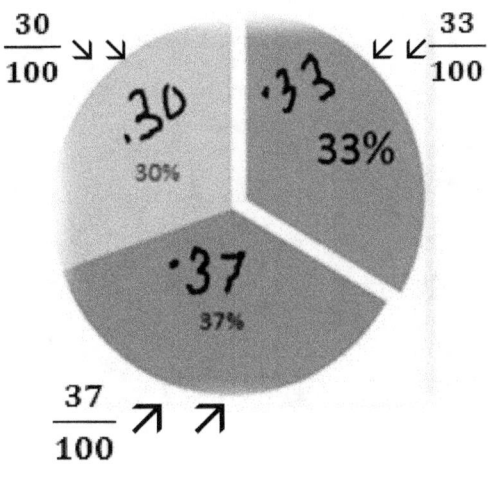

$.30 = 30\% = \frac{30}{100}$

$.33 = .33\% = \frac{33}{100}$

$.37 = .37\% = \frac{37}{100}$

100 = 100% = Whole circle represents 1 piece of object. The whole circle has three colored parts in which representing respectively 37% brown color; 33% blue; 30% green.

CBEST Math Section:
California Basic Educational Skills Test

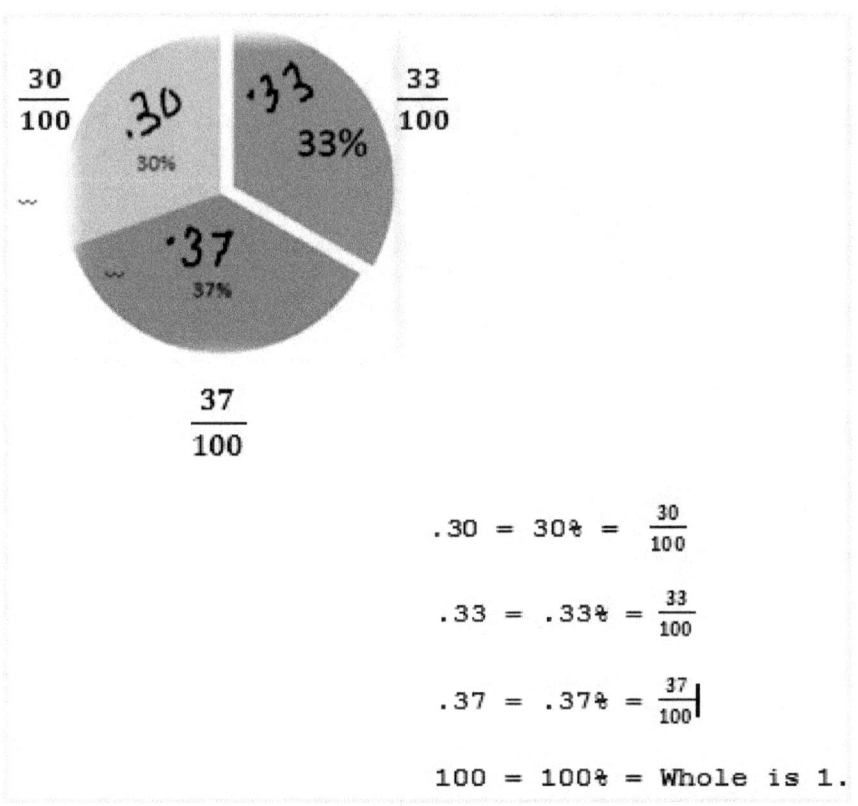

$$.30 = 30\% = \frac{30}{100}$$

$$.33 = .33\% = \frac{33}{100}$$

$$.37 = .37\% = \frac{37}{100}$$

$$100 = 100\% = \text{Whole is 1.}$$

Most frequently used percent (%) and decimal digit conversation charts.

CBEST Math Section:
California Basic Educational Skills Test

Percentage conversion to Decimal point:

1% = .01
5% = .05
10% = .10
20% = .20
30% = .30
40% = .40
50% = .50

60% = .6
70% = .7
75% = .75
80% = .8
85% = .85
90% = .9
100% = 1.00

CBEST Math Section:
California Basic Educational Skills Test

Second example is a giant rectangle object:

It consists of small (10*10)=100 boxes in which 25 boxes are colored. The 25 boxes also represents 25% of area colored. 25% boxes are colored and remaining 75 boxes represent 75% colored.

Again, total area of that rectangle object contains (10x10)= 100 boxes makes a rectangle. The object is represent one unit that subdivided 100 small boxes as a 100%.

CBEST Math Section:
California Basic Educational Skills Test

Measurement:

Colored area is (5x5) = 25 boxes

Total rectangle consists of 100 small boxes.

What is percentage represents of colored boxes area?

Change fraction to decimal:

Colored area 25 small boxes,

$$\text{Fraction} = \frac{\cancel{25}}{\cancel{100}} \frac{1}{4} = \frac{1}{4} \text{ (lowest terms)}$$

Change decimal to fraction:

$$\text{Fraction} = \frac{1}{4} \text{ (multiply by 100)}$$

Or $\frac{1}{4}$ x 100 (Change Decimal hundredth)

$$\text{Or } \frac{1}{4} \overline{)10} \,(.25$$
$$\underline{8}$$
$$20$$
$$\underline{20}$$
$$00$$

Fraction $\dfrac{1}{4}$ becomes decimal **.25**

Change numbers in various forms:

Percent to fraction; percent to decimal, decimal to fraction, and decimal to percent

Percentage formula,

$$\text{Percent} = \frac{Colored\ area}{Total\ Ractangle\ Area} = \frac{25}{100} =$$

$$\frac{1}{4} = 0.25$$

$$0.25 \times 100 = 25\%$$

To change a fraction to a percent follows the step # through Step #3 noted hereafter.

Let's take a fraction number: $\frac{1}{4}$

Step #1: Multiply by 100.

(The process allows fraction change to decimal, Where $\frac{1}{4}$ fraction to be changed into decimal point. Because, decimal define a fraction whose denominator is a power of ten and whose numerator is expressed by figures placed to the right of a decimal point.

For instance, $(\frac{1}{4} \times 100) = \frac{100}{4}$

Step #2: Divide by denominator.

For instance, $\dfrac{100}{4} = \dfrac{100}{4} = \dfrac{50}{2} = \dfrac{25}{1}$

(lowest terms)

Step 3: <u>Insert a percentage</u> (%) sign in front of outcome number.

Recall step#3, $\dfrac{25}{1} = 25 = 25\ \%$ (place percent sign)

Second similar Example:

Fraction number $\dfrac{1}{2}$

Change to decimal needs multiply 100

$\dfrac{1}{2} \times 100 = \dfrac{100}{2} = \dfrac{50}{1}$ (change to lowest terms

Recall $\dfrac{50}{1}$ from the above lowest terms, which then becomes 50

Place the percent sign with it, i.e. 50%.

CBEST Math Section:
California Basic Educational Skills Test

Problem #1:

How do you change *a percent to fraction?*

Well, take a percentage number like 50%.

It takes two steps procedure given bellows.

Step #1:

Take out the percent sign.

For instance 50% to 50

Step #2:

Divide the percent by 100.

For instance $\frac{50}{100} = \frac{1}{2}$ (lowest terms)

So, 50% becomes fraction $\frac{1}{2}$

CBEST Math Section:
California Basic Educational Skills Test

Problem #2

How do you change a <u>fraction to decimal</u>?

For example, take a common fraction like this:

$$\frac{11}{20} = 20\overline{)11.0}(.55$$
$$\underline{10\,0}$$
$$100$$
$$\underline{100}$$
$$x\,x$$

Fraction $\frac{11}{20}$ becomes **decimal 0.55**

Alternate Method:

Fraction $\frac{11}{17}$ continuing)_

$$17\overline{)110}(.647xx \text{ (x represents}$$
$$\underline{102}$$
$$80$$
$$\underline{68}$$
$$120$$
$$\underline{119}$$
$$1x$$

Fraction $\frac{11}{17}$ becomes decimal **0.647...**

CBEST Math Section:
California Basic Educational Skills Test

Problem #3:

How do you change a *decimal to a fraction?*

Well, take any decimal point number to change into fraction, let's take .55

It takes two steps procedure given bellows.

Step #1:

Move the decimal point two places to the right, for example, 0.55 is a decimal number. Its place value 0.5 tenth place, and 0.05 hundredth places, which denoted two places after the decimal sign.

Step #2:

Put the decimal number over the 100. Look at the following example,

$\frac{55}{100}$, which means 55 out of 100.

$\frac{11}{20}$ (Lowest terms)

CBEST Math Section:
California Basic Educational Skills Test

Decimal 0.55 becomes fraction $\frac{11}{20}$

Few examples of conversion data from decimals to fraction and their lowest terms

Decimal ↓	Fraction ↓	Lowest Terms ↓
.6	$\frac{6}{10}$	$\frac{3}{5}$
.7	$\frac{7}{10}$	N/A
.8	$\frac{8}{10}$	$\frac{4}{5}$
.9	$\frac{9}{10}$	N/A
.09	$\frac{9}{100}$	N/A
.009	$\frac{9}{1000}$	N/A
.0009	$\frac{9}{10000}$	N/A
.00009	$\frac{9}{10000}$	N/A

Problem 3

How do you finding percent of a number?

To determine percentage (%) of a number. it needs to change the percent to a decimal or to a fraction. It then multiplies outcome is the answer.

For example:

15% of 60. It means $0.15 * 60 = 9$

Remember that 15% of 60, it noticed the preposition "of" served as a *multiplying* symbol. Thus, the preposition "of" is meant a multiplication purpose.

Again, *of* is used substitute of multiplication function. It is one of the math word jargon interpretation that solve math equation.

CBEST Math Section:
California Basic Educational Skills Test

CBEST Math Test Watch!

Learn from Examples:

Problem 1:

What is 25% of 80?

Solution:

$$25\% = \frac{25}{100}$$

(25 percentages means that it a 25 parts out of 100 parts)

Therefore, $\frac{25}{100} \times 80 = \frac{2000}{100} = 20$ (lowest terms)

So, 20 is the answer

The number calculations,

(25% of 80) becomes 20.

CBEST Math Section:
California Basic Educational Skills Test

Problem 2:

What is $\frac{3}{4}$% of 20?

$\frac{3}{4}$% of 20 or $(\frac{\frac{3}{4}}{100}) \times 20$

Or $(\frac{\frac{3}{4}}{\frac{100}{1}}) \times 20$ (3/4 divide 100)

$= (\frac{3}{4} \times \frac{1}{100}) \times 20$ (simplify fractions)

Or $\frac{3}{400} \times 20$ (Lower terms)

$= \frac{3}{20}$ (fraction value, ¾ of 20)

Or $\frac{3}{20} =$

Or $\frac{3}{20}$ = $20\overline{)30}\,(.15$
$\phantom{20\overline{)}}\underline{20}$
$\phantom{20\overline{)}}100$
$\phantom{20\overline{)}}\underline{100}$
$\phantom{20\overline{)1}}00$

The value, $\frac{3}{4}$% of 20, becomes 0.15

CBEST Math Section:
California Basic Educational Skills Test

Arithmetic Applications Word into Sign:

Turn Word-for-Word into various equations signs

For example,

Word substitutes:

"is" = equal to

"of" is substitute multiplying sign X

"what" is used to find unknown value/number/answer etc.

Example of word's substation:

Example #1:

20 is what 90%? It can substitute like this:

20 is what percent of 90?

We can say, $90\% = \dfrac{90}{100}$

$20 \times \dfrac{90}{100} = \dfrac{180}{100} = 18$ (lowest terms)

The 18 outcomes turns into percent sign, **so it is 18%**

Example #2:

What is 10% of 45%?

$$\frac{10}{100} \times 45 = \frac{1}{10} \times 45 = \frac{45}{10} = 4.5$$

The number 4.5 change to percent, just add percent sign after the number; therefore, it is 4.5%.

10% of 45 is 4.5%

The answer is 4.5%

CBEST Math Section:
California Basic Educational Skills Test

CBEST WATCH: Practice Test Question #2

It involves Word/phrase Problem solving such as PERCENTAGE (%)

2. Use the table below to answer the question that follows.

Section	Total Number of Questions	Number of Questions Correctly Answered
Algebra	20	17
Trigonometry	15	11
Geometry	25	20

On the three sections of a math test, a student correctly answered the number of questions shown in the table above. What percent of the questions on the entire test did the student answer correctly?

A. 20%

B. 48%

C. 75%

D. 80%

E. 96%

Ref: http://www.ctcexams.nesinc.com/PDF/CBEST_OPT_Math.pdf

The question inquiries the keyword percentage (%) between total number of questions of 3 sections and numbers of questions correctly answered by the student.

CBEST Math Section:
California Basic Educational Skills Test

The question given Algebra, Trigonometry, and Geometry are 3 sections. It is noted bellow.

Section	Total Number of Questions	Number of Questions Correctly Answered
Algebra	20	17
Trigonometry	15	11
Geometry	25	20
Total	60	48

Student correctly answered 48 out of 60 questions. So, it needs Percentage (%) formula.

$$\text{Percent \%} = \frac{48}{60} \text{ (Reduce to lower terms)}$$

$$= \frac{24}{30} = \frac{12}{15} = \frac{4}{5}$$

$$= \frac{4}{5} = \frac{\square}{\square}$$

= 0.80 (multiply 100 to change percentage).

= 80%   The answer is D.

CBEST Math Section:
California Basic Educational Skills Test

How do you find out percent change (increase or decrease results)?

To find out the percentage change value, it uses percentage change formula, which makes % increase or decrease value.

$$\text{Percentage change} = \frac{Profit\ or\ Loss\ \$Value}{Initial\ Investment\ \$\ Value} \times 100$$

Change value = Starting value minus (-) Ending value, for example, if you have sold $500.00 goods, which have bought $300.00.

Change value = $500.00 -$300.00 = $200

Profit value is $200.00

Profit is $200.00. It creates question what is the percent increase. Look at the example hereafter.

Learn from Examples:

Example #1:

A business-woman invested $300.00, which finally (principle plus profit) retuned $500.00. What is the percentage increase?

Solution:

Change value = $500.00 - $300.00 = $200.00

Profit = $200.00

Percentage change = $\frac{\$200}{\$300} = \frac{2}{3} = 0.667 = 67\%$

Total Profit percentage 67%.

Example #2:

Mr. John invested $300.00, which finally retuned $200.00. What is the percentage decrease?

Solution:

Change value = $300.00 - $200.00 = $100.00

Investment Loss = $100.00

Percentage change = $\frac{\$100}{\$300} = \frac{1}{3} = 0.333 = 33\%$

Total investment change negative percentage -33%.

CBEST Math Section:
California Basic Educational Skills Test

Ratio & Proportion:

What is ratio?

A ratio is actually come from fraction, for exampe, 3/2 is a fraction, which represent 3:2. This fraction has the quantitative relation between two numbers i.e. 3 and 2 that showing the numbers of times one value contains within other number – quotient.

Quotient of the two that indicates how many time 1st number contains the 2nd number. For example, one busket contains 8 blue marble and 6 red marble. In this case, the ratio between blue and red marble is 8:6 =4:3 (simplified lowest).

Further, it can be represented fraction :

The ratio blue to red: 4/3 = 4:3

The ratio red to blue: $3/4$ =3:4

Colon : sign represents ratio

The ratio of blue marble to total number of marble in the busket: 8/14=4/7 (can't lower)

Where total 8 blue + 6 red marbles =14 that ratio with blue

So, 8/14 that is lowest fraction is 4/7. It ratios 4:7.

CBEST Math Section:
California Basic Educational Skills Test

Proportion:

What is proportion?

A proportion is an equation that stating two ratios are equal. It means that if ratio is 4:7 it can rewrite 4/7 or raise it proportion multiplying eqal number both parts, for example, 2 (4/7) = 8/14.

Another example is if ration 5:10 = 5/10 = ½

Let's say, Ms Simson's class contains 15 boys and 12 girls. What is ratio between boys and girls?

Boys : Girls = 15:12 simplifying it that comes 5:4 (dividing by 3 both side of the equation.

Simply we can say Mrs Simosn's class boys and girls ratio 5:4.

The 5:4 ratio does not represent actual boys 15, girls 12 in the class room. The ratio 5:4 can be represented higher number proportion of students of both sex, for example, boys 30 and girls 24. Why's that?

Becasue 30/24 = simplifying became 5:4.

Proportion does not hold constant figures. Its applications are everywhere, for example, a map shown that the distance between John's home and school is 4" (four inch). It foot note says 1" = 1000 yards. What is the actual physical distance that John travel every day?

we know map's distance 4" John travel each day and 1" = 1000 yards. We can use this information to calculate actual distance between home to school and school to home that John travel everyday.

The distance 4" = ?

Where 1" = 1000 Yards

4" x 1000 = 4000 yards oneway

4000 x 2 = 8000 yards.

Alternate method to solve it as follows:

Where 1" represents 1000 yards, total map's distance 4"

It can represent ratio 1:4; thus, proportion multiplying both side by 1000, which becomes 1000: 4000

So, the 4" = 4000 Yards, which is one way travel.

Up and down travel (2x4000)= 8000 yards.

Prime Numbers:

Prime number, which can be evenly divided by only itself and 1, for instance, 11, 13, 17, and 19 are prime numbers. Because they can be divided by itself i.e. 13/13 = 1, 17/17 =1, 19/19=1 and they also divided by 1 (one), for example, 13/1=13, 17/1=17, 19/1=19; therefore, they are called prime numbers.

On the other hand, 21 is **not** prime number because it can be divided by 3 and 7.

Even Prime Number

Number 2 consider even prime number. It can be divided by 2, which results 1. Other prime numbers are 2, 3, 5,7,23, and 29.

NonPrime Number:

Nonprime Numbers are integer, which not prime numbers i.e. 1 and 0 (zero).

Positive nonprime and nonnegative numbers are listed hereafter.

CBEST Math Section:
California Basic Educational Skills Test

Positive nonprime numbers

Positive nonprime numbers: the unit 1 and the primes.

A018252 The nonprime numbers (the unit 1 together with the composite numbers, A002808.)

{1, 4, 6, 8, 9, 10, 12, 14, 15, 16, 18, 20, 21, 22, 24, 25, 26, 27, 28, 30, 32, 33, 34, 35, 36, 38, 39, 40, 42, 44, 45, 46, 48, 49, 50, 51, 52, 54, 55, 56, 57, 58, 60, 62, 63, ...}

Nonnegative nonprime numbers

Nonnegative nonprime numbers: 0, the unit 1 and the primes.

A141468 Zero together with the nonprime numbers A018252.

{0, 1, 4, 6, 8, 9, 10, 12, 14, 15, 16, 18, 20, 21, 22, 24, 25, 26, 27, 28, 30, 32, 33, 34, 35, 36, 38, 39, 40, 42, 44, 45, 46, 48, 49, 50, 51, 52, 54, 55, 56, 57, 58, 60, 62, 63, ...}

Typical listing collection from Oeis.org Wiki

Reference: http://oeis.org/wiki/Nonprime_numbers

W

CBEST Math Section:
California Basic Educational Skills Test

Which of the following is a prime number?

A. 97

B. 0

C. 1.5

D. 57

E. 1

The answere is A. 97

CBEST Math Section:
California Basic Educational Skills Test

FACTORING:

Whole Numbers' Involement of Factoring:

Factoring, Divisors, and Multiples are three important things that addressed in the CBEST Math.

These are not whole numbers:
1. Fractions
2. Decimals
3. Negative numbers.

What is factoring?

A number that can be broken into two numbers so it can be multiplied together to obtain the original numbers. For example, 36 = 9 * 4. So, both numbers can be further makes factoring. They express: factoring 9, it esults: 3*3; factoring 4, it results: 2*2

Factoring prime numbers are 2,3,5,7,11,13,17,19, which all of them can make foctoring with the the number 1, for exmple, 19*1= 19; 2*1=2 etc.

What is LCM?

LCM stands for Lowest Common Multiple .Lowest, it interchangeably says Least or smallest number. It needs when we calculate and use prime factorization of an equation, for example, the number 12 has factors; 2*2*3.

CBEST Math Section:
California Basic Educational Skills Test

The number 72 that contains serveral factors like this:

72 = 9*8=(3*3)*(2*4)= 3*3*2*(2*2)= 3*3*2*2*2

The number 72 contains:

   The number 2 appeared 4 times

   The number 3 appear 2 times.

The number 48 that contains serveral factors like this:

48 = 8*6=(2*4)*(2*3)= 2*(2*2)*2*3= 2*2*2*2*3

The number 48 contains:

   The number 2 appeared 4 times

   The number 3 appear 1 time

Finding the LCM of 48 and 72 is to follow the following steps describe hereafter.

How many common factors are there? The numbers are 72 and 48.

It calculation method is that how many times each prime factor appeared most, 72 gives 4 time 2 and 48 also gives 4 times 2. In this senerio we conclude that both (72 and 48) have common number 2 appeared 4 times, which becomes common both it means (2*2*2*2) =16

Next, calculation is 72 and 48 in which have also common 3. It appeared 3 number two times within 72 and 1 time within 48. It means we take (3*3)=9

Now, common LCM of 72 and 48 is (16*9)=114

CBEST Math Section:
California Basic Educational Skills Test

What is GCF?

GCF stands for Greatest Common Factor. It also needs when we calculate and use prime factorization of an equation. However, it contains uncommon number. The method is to multiply the common factors that appear most of the time from the pair of numbers it (GCF) given.

GCF is the same as LCD, however, one step is different when pairs have uncommon numbers, for example, GCF of 60 and 54 would be as follows:

$60 = 10*6=(2*5)*(2*3)=2*2*3*5$, it give 2 times 2 and one time 3 and one time 5.

$54 = 9*6 = (3*3)*(2*3)=3*3*3*2$, it given 3 times 3 and one time 2

Next,

$60 = 2*2*3*5$, it contains 5 that is uncomon number.

$54 = 2*3*3*3$

From the above calculation, we determine that 3 common number appeared once and 2 also appeared once but 5 is not common. So, we can concluded that their (60 and 54) GCF is $(2*3)=6$

What are the LCM and GCF of 30, 36, 48, and 72?

For LCM

$30=6*5=2*3*5$, 5 appeared once that is candidate #3

CBEST Math Section:
California Basic Educational Skills Test

36=6*6=(2*3)*(2*3)=2*2*3*3, two times 3s, it appeared 2nd most numbers #2

48= 2*2*2*2*3, four times 2s that most times appeared, it is the candidate #1

72= 2*2*2*3*3, two times 3s, it candidate #2

Number four times 2s * two times 3s* one time 5 =LCM

LCM = #1*#2*#3 = (2*2*2*2)*(3*3)*(5)=16*9*5=720

For GCF

30=6*5=2*3*5, number 5 appeared once #1

36=6*6=(2*3)*(2*3)=2*2*3*3, number 3 & 2 twice #3

48= 2*2*2*2*3, number 2 four times, 3 once #2

72= 2*2*2*3*3, number 2 thrice and 3 twice #2

2 and 3 are common number of 36, 48, and 72.

GCF= #1*#2*#3= 5*3*2=30

LCM and GMF are two components that useful in the operation with fractions.

Most often in the classroom LCM practices are:84, 96; 48, 52; 8, 14, 24; theirs LCM respecitively are 12, 4, and 168.

What is composite numbers?

A composite number is a positive integer in which has at least one positive divisor that other than one or the number itself.

Alternately, a composite number is any positive integer number that is not a prime number and that number is greater than 1.

For example, as we know that 6 is not a prime number. Why's? 6 can be divided by 1,2,3. Therefore, it is composite number. However, some whole number can be divided evenly but some are not.

A composite number can be divided up evely while a prime number can't, for example, 51 is a composite number because $51 \div 3 = 17$, thus 3, 17 factors of 51.

What is prime number?

A prime number can be divided evely by 1 or itself and also it to be a whole number that is greater than 1.

For example, 62 can't factor becuase 62 when divided by the following numbers 2,3,5,7 that produced reminder except by 1. It is therefore prime number and not possible to make foctors. 42 has factor of 6 although it is not a prime number see listing of noprime number chart.

What is quotient?

A quotient is an outcome when you dividing one number by another number. It can be expressed like this:

Dividend divided by divisor = quotient

or dividend $\div$ divisor = quotient, for example, $20 \div 4 = 5$, it outcomes 5, it is called quotient.

CBEST Math Section:
California Basic Educational Skills Test

What is divisor?

A number that has no remainder when it divides by another number. For example, 15/3 = 5 while 14/3=4.6666, has remainder i.e., 0.666 therefore it is not a qualify divisor.

What is multiply?

A multiply is, for example, 5*3=15, it can also repeated additon, for exmple, 5+5+5=15.

Length Measurement:

Traditional Method

1 foot (ft) = 12" inches (in)
1 yard (yd) = 3' or (3 x 12) = 36"
1 mile (mi) = 1760 yards (1760 x 3) = 5280'
Symbol Index:
Foot/feet/ft symbol is look like this (')
Inch/inches symbol is look like this (")

Traditional Vs. Metric System

Inch is a unit of linear measurement that equal to one twelfth (1/12) of a foot, which equivalent to 2.54 cm (centimeter).

CBEST Math Section:
California Basic Educational Skills Test

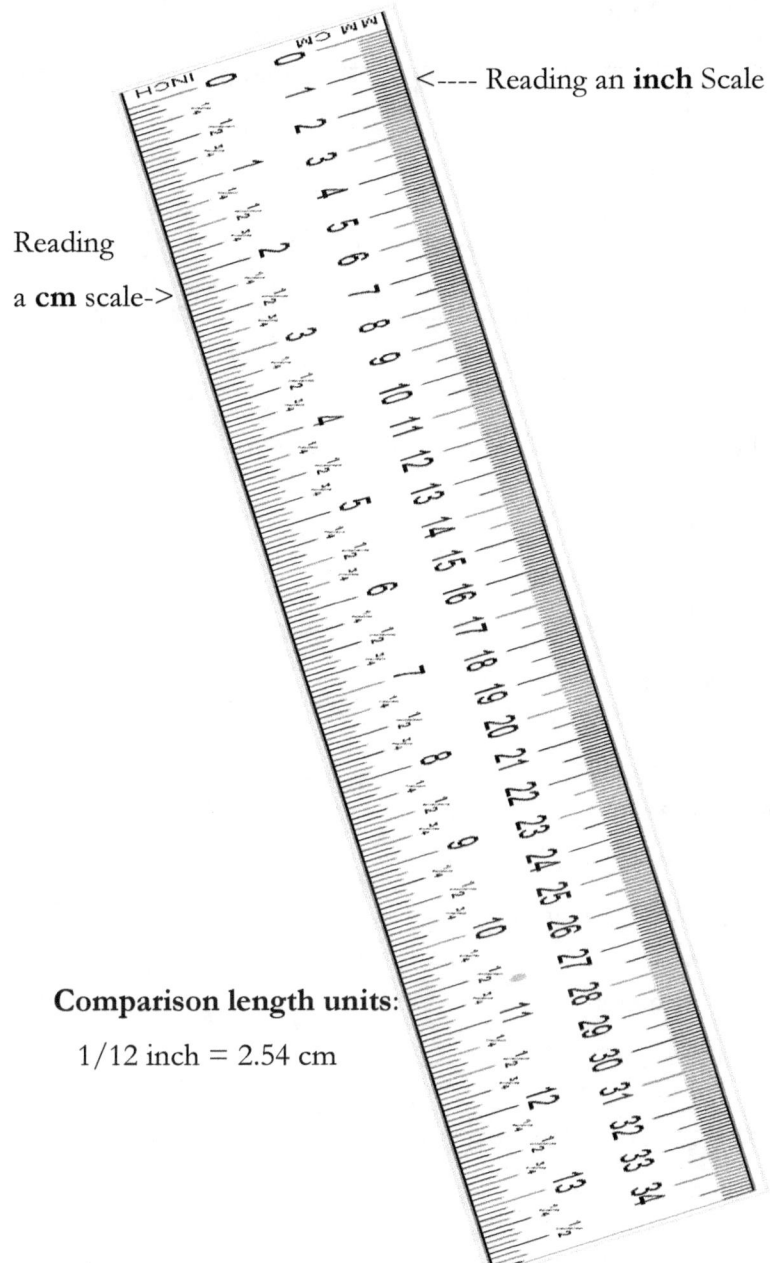

<---- Reading an **inch** Scale

Reading a **cm** scale->

Comparison length units:
1/12 inch = 2.54 cm

Typical Actual size measurement Scale

CBEST Math Section:
California Basic Educational Skills Test

2.54 cm = 1 inch

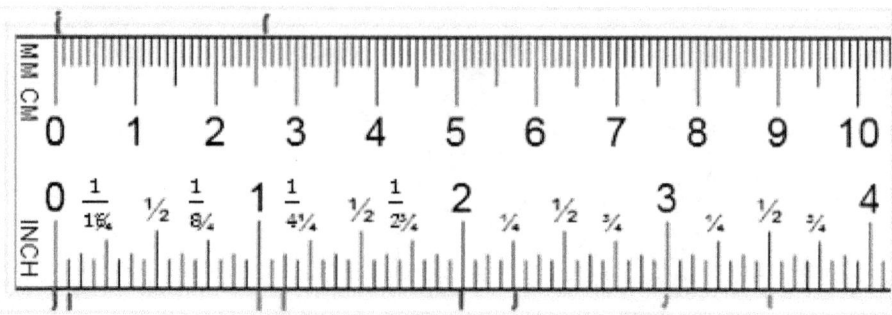

1 inch divided into 16 parts that 1 part represents $\frac{1}{16}$

2 parts represent $\frac{1}{8}$

4 parts represent $\frac{1}{4}$

8 parts represent $\frac{1}{2}$

Inch

1 inch = 2.54 mm

Feet

1 foot = $\frac{12}{39.37}$ or 0.3048 meter

1 foot = $\frac{5}{16}$ meter

Yard

1 yard = $\frac{3600}{3937}$ or 0.9144 incj

CBEST Math Section:
California Basic Educational Skills Test

Weight Measurement

1 gm (gram) = 1000 mg (milligrams)

1 kg (kilogram) = 1000 gm

1 t (metric ton) = 1000 kg

Gram abbreviated = gm

Kilogram = kg

Metric ton = t

Capacity Measurement Liquid.

1 liter = 1000 milliliters

1 kiloliter = 1000 meters

L, mL, kL are abbreviated respectively Liter, Milliliter, and Kiloliter.

Time unit

60 seconds = 1 minute

60 minute = 1 hour

24 hours = 1 day

7 days = 1 week

365 days = 1 year;

Leap year = 366 days

52 weeks = 1 year

1 decade = 10 years

1 century = 100 years

CBEST Math Section:
California Basic Educational Skills Test

Time unit, both metric and traditional systems time measures same way.

Temperature unit

Celsius (°C)
0°C = 32°F
1°C = 1.8°F
100°C = 180°F

Fahrenheit (°F)
1°F = 0.556°C
180°F = 100°C

CBEST Math Question Examples:

Use the scale below to answer the question that follow.

The CBEST Math section may ask qustion similar to this:

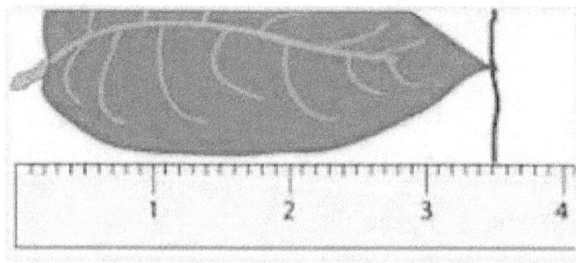

What i s lenght of the above object?

CBEST Math Section:
California Basic Educational Skills Test

Find the lenght of an object, for instance, total length of an object (measure solid cm portion) plus fractional parts of a cm, if any.

Note that 1 cm divided into 10 parts, which 1 part is known as 1 millimeter.

The scale is centimeter or cm.
Look at the object given below.

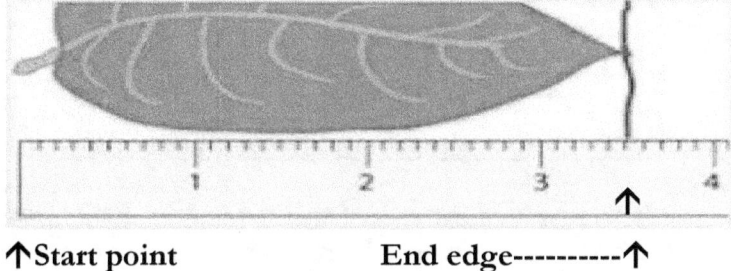

↑Start point   End edge----------↑

The leaf illustrated measures 3 solid centimeter (cm) plus 5 parts of a cm or $\frac{5}{8}$ parts of a cm). It measures: $3\frac{5}{8}$ or 3.5 cm

CBEST Math Section:
California Basic Educational Skills Test

CBEST WATCH:

QUESTION #3:

3. Use the diagram below to answer the question that follows.

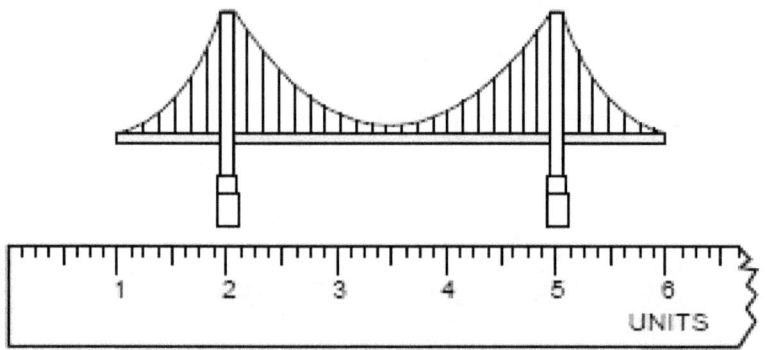

If the actual length of the bridge is 4200 feet, then what is the scale of the diagram of the bridge?

A.   1 unit = 700 feet

B.   1 unit = 763.6 feet

C.   1 unit = 840 feet

D.   1 unit = 933.3 feet

E.   1 unit = 1050 feet

Solution:

The question provided the actual length of the bridge is 4200 ft.

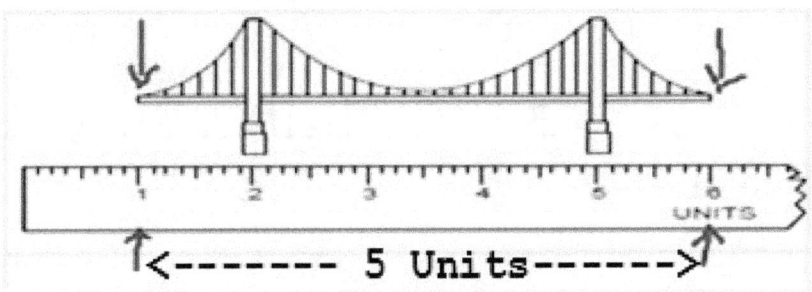

This question needs carefully reading the scale's unit counts. The bridge measurement started 1 and eded 6. Therefore, it lenghts (6-1)= 5 units.

Now, the scale reading of the bridge is 5 units, which representing 4200 ft.

Therefore, $\dfrac{4200}{5} = 840$ feet

The answer choice is **C**.

CBEST Math Section:
California Basic Educational Skills Test

Question #01:

Learning objective: Reading franction of an inch

How far is point B from point A?

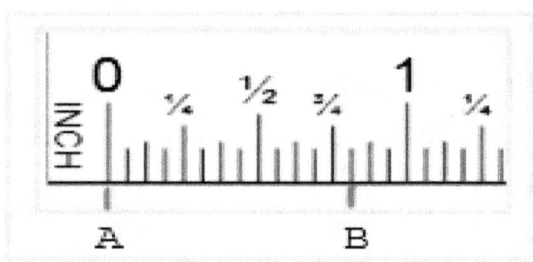

A. $\frac{11}{16}$

B. $\frac{3}{4}$

C. $\frac{13}{16}$

D. $\frac{7}{8}$

E. $\frac{15}{16}$

Reading scale ability is the key to figure out the answer of this type question.

Point A is the begging point and it ends at 13 parts of an inch. As we know, 1 inch is divided by 16 parts; therefore, it is $\frac{13}{16}$ of an inch.

The answer is c.

Perimeter and Areas:

What is perimeter?

Perimeter measures surrounding of an object like a trangle or a polygon. It calculates the sum of the length of all sides the object. For example, if an object is rectangle then it perimeter calculation uses this this formula:

(Lenght + Width) x 2 = 2L + 2W

width

Length

Area of Rectangle formula:

Area = L * W

Square Area

An object is square that formula:

S represents 1 side. It has four equal sides, which formula is s^2

s^2

s= 4" it perimeter then 4^2 = 4*4=16"

Traigle Area:

The area of a triangle formula

where A = area;   b = base; h = height

Area, A= ½ (b*h)

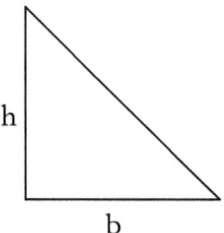

Circles:

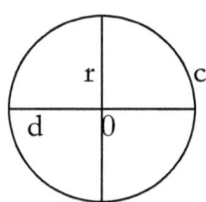

- Diameter = d, a line starts one side of a circle and passess through the center to the other on the circle arc.
- Radius = r, a line that starts at the center of a circle and ended on the circle arc.
- Circumference = c, the length of the circle arc
- c = πd (pi *d) = 2pi*r
- pi = 3.14 = 22/7
- r = ½ d
- Area of a circle, A =pi *r^2

CBEST Math Section:
California Basic Educational Skills Test

Perimeter

Add up all sides of an object, which makes the perimeter.

CBEST MATH WATCH! QUESTION #7: Perimeter

7. Use the diagram below to answer the question that follows.

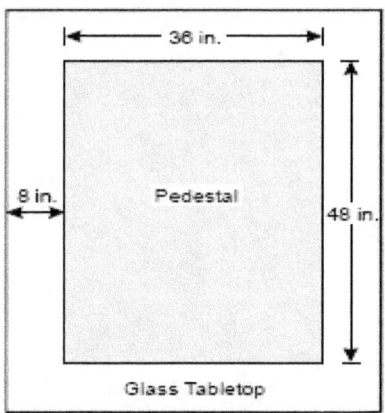

A glass tabletop is supported by a rectangular pedestal. If the tabletop is 8 inches wider than the pedestal on each side, what is the perimeter of the glass tabletop?

A. 92 inches

B. 116 inches

C. 176 inches

D. 184 inches

E. 232 inches

Rectangle pedestal measurements are given:

Length is 48 inches

Wide is 30 inches

Tabletop each side is extended 8 inches.

Total length of tabletop is (48"+8"+8") =64"

Total wide tabletop is

(36"+8"+8") =52"

What is perimeter of the glass tabletop?

The keyword is "Perimeter", which means 2 times length + 2 times wide.

So, (2x64) + (52x2)

= 128 + 104

= 232 inches

The choice is **E**.

CBEST Math Section:
California Basic Educational Skills Test

Question #02:

Learning Objective: Fraction

Illustration of fraction of an inch

How far is point **C** from left end of the scale?

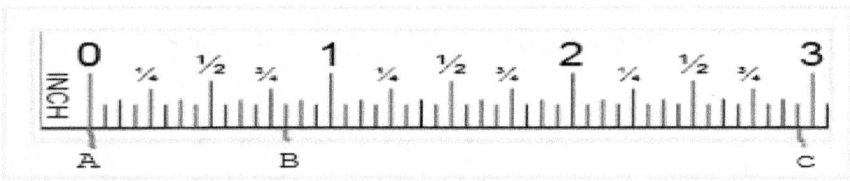

A. $2\frac{11}{16}$

B. $2\frac{3}{4}$

C. $2\frac{13}{16}$

D. $2\frac{7}{8}$

E. $2\frac{15}{16}$

Answer analysis:

First notice that point C is between 2 and 3 inch scale marked. Secondly, the C point is located between 2" and 3" marked. So, we know it is 2" + how many parts beyond the 2" marked.

Third, it counted 15 parts out of 16 parts.

Therefore it measures $2\frac{15}{16}$.

The answer is E.

CBEST Math Section:
California Basic Educational Skills Test

Learning Objectives:

Simplify the problem; combining terms; performing simple operations.

Question #03:

Which of the followig is between $\frac{1}{4}$ and 0.335?

A 0.029

B 0.29

C 2.90

D 0.37

E 0.41

Convert the value 1/4 into decimal is the only the way to find the middle number betwween ¼ and .3375, which is 0.25 or 1/4.

$\frac{1}{4} = 0.25,$

SO, alternately we can say,

which of the following is between 0.25 and 0.335?

Thus the result is B.

CBEST Math Section:
California Basic Educational Skills Test

Question #04:

Which of the following is equal to $\frac{1}{4}$ of 0.02%?

    A 0.5

    B 0.05

    C 0.005

    D 0.0005

    E 0.00005

To solve this problem, it needs simplifying ¼ first, which is 0.25.

So, 0.25 of .02% = 0.25 x 0.02%

or $(0.25 \times \frac{02}{100})$

or (0.25 x 0.0002)

or 0.00005

The choice is E.

CBEST Math Section:
California Basic Educational Skills Test

Question #05:

The mixed number $1\frac{7}{9}$ may be placed between which of the following nearest franctions?

A. $\frac{15}{9}$ and $\frac{19}{9}$

B. $\frac{13}{9}$ and $\frac{15}{9}$

C. $\frac{7}{9}$ and $\frac{11}{9}$

D. $\frac{11}{9}$ and $\frac{7}{9}$

E. $\frac{5}{3}$ and $\frac{17}{9}$

Understand the mixed number is very important to solve this type of question. One of the ways it could be properly identified is break-down this number $1\frac{7}{9}$.

Let's see convert it into decimal point value.

The number is $1\frac{7}{9} = \frac{16}{9} = 1.78$

Next step look at the number 1.78, which is fit between which of the fractions that are provided choice A though E. Let's find A, which provided $\frac{15}{9}$ and $\frac{19}{9}$.

Thus, $\frac{15}{9} = 1.67$ and $\frac{19}{9} = 2.11$

We can say, 1.67 and 2.11 instead $\frac{15}{9}$ and $\frac{19}{9}$

Similarly their decimal value are converted below.

$16/9 = 1.78$

A. $15/9 = 1.67$
$19/9 = 2.11$

B. $13/9 = 1.4$
$15/9 = 1.67$

C. $7/9 = 0.78$
$11/9 = 1.22$

D. $11/9 = 1.22$
$7/9 = 0.78$

E. $5/3 = 1.67$
$17/9 = 1.89$

Choice A starts 1.67 and ended 2.11; The ending 2.11 is far from 1.78 that's why it is not the nearest.
The best choice E, which is the nearest number between 1.67 and 1.89.

CBEST Math Section:
California Basic Educational Skills Test

Weight Measurement

1 lb = 16 oz.

1 T = 2000 lbs

Symbol Index:

Pound abbreviated = lb. (plural lbs)

Ounce abbreviated = oz.

Ton abbreviated = T

CBEST MATH TEST WATCH!

QUESTION #4: Weight Measurement

Ref:

http://www.ctcexams.nesinc.com/PDF/CBEST_OPT_Math.pdf

---

4. Which of the following is the most appropriate unit for expressing the weight of a pencil?

    A. pounds

    B. ounces

    C. quarts

    D. pints

    E. tons

The question provided 5 choices in which choice A and choice B are related with weight. However, choice A (pounds) could not be a pencil weight. Remaining choices C (quarts) & choice D (Pints) are

CBEST Math Section:
California Basic Educational Skills Test

using to measure liquid. Finally, choice E is 'Tons'. Therefore, choice E "Tons" is very big weight measuring unit, which can't be a pencil weight. The best choice is **B**.

Measurement Units: Weight

Pound is a unit of weight that
   equal to 16 ounces (oz).

Ounce is a unit of weight of
   one-sixteenth ($\frac{1}{16}$) of a pound.

Quart is a liquid capacity equal to
   a quarter of a gallon or two
   pints, equivalent in approximately
   1.13 liter (UK measurement system)
   and 0.94 liter (US to approximately.

Pint is a unit of liquid or dry capacity
   to one-eighth ($\frac{1}{8}$) of a gallon.
   In UK system, 1 pint=0.568 liter
   & US system, 1=0.551 liter).

Ton is a unit of weight equal to 2240 lb.
   It's also a unit of gross internal
   capacity that equivalent to
   100 cubic feet or 2.83 cubic meters.

CBEST Math Section:
California Basic Educational Skills Test

Liquid Measurement:

1. Capacity Measurement (liquid).

    1 c (cup) = 8 fl oz (fluid ounce)

    1 pt (pint) = 2 cups

    1 qt (quart) = 2 pints

    1 gal (gallon) = 4 quarts

    c, pt, qt, and gal are abbreviated respectively cup, pint(s), quarter(s), gallon(s).

2. Time Measurement.

    1 min = 60 sec

    1 hr = 60 min

    1 day = 24 hrs

    1 week = 7 days

    1 year = 365 days

    Symbol Index

    Hour abbreviated = hr or hrs (pl)

    Week abbreviated = wk or wks (pl)

    Year abbreviated = yr or yrs (pl)

CBEST Math Section:
California Basic Educational Skills Test

Metric System Measurement Unit:

Metric Method or CGS System

What is CGS system?

CGS stands for **C**entimeter, **G**ram, and **S**econd.

They are lowest measurement unit representation of CGS system

It measures Length in Centimeter (cm)

Weight in Gram (gm)

Time in Second (sec)

Meter:

1 meter (m) = 100 centimeter (cm).

1 (cm) = 10 millimeter (mm)

1 (mm) = 0.01 cm or 0.001 meter

Symbol index:
Centimeter abbreviated = cm
Meter abbreviated = m
Kilometer abbreviated = km

Relation between mile and KM

Mile is a unit of linear measurement.

1 mile = 1,760 yards or 1.609 KM

CBEST Math Section:
California Basic Educational Skills Test

Analysis of a typical landscape application using mile measurement:

The diagram below showed an area of 20 miles long and 14 miles wide area. It's also fact that each box area represents (2x2) = 4 squared-miles. The total area is a rectangle shape area of (20x14) =280 square-miles where a lake is situated.

The typical CBEST test is:

What is the Clear Lake area of the lake shown below?

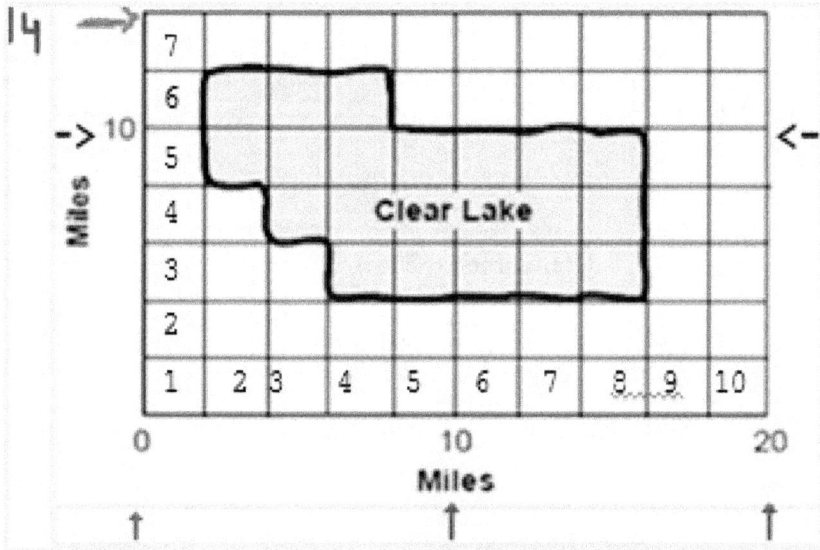

Area measurement needs mile or kilometer. In the US, landscape or distance measurement uses mile unit. Again, how do you figure out the <u>area</u> of the lake in square mile? It is a good question, isn't it?

CBEST Math Section:
California Basic Educational Skills Test

First step:

When you solve this type of question then it provides sufficient information such as measurement indication. So, it needs concentration on reading the diagram such as what is the area of a single box. Again, the diagram has horizontal line, which starts 0 (zero) and ends 20; however, it contains 10 division horizontally that information give you the clue to figure out the vertical distances. It means each box length horizontally 2 miles and vertically 2 miles wide. Aren't it? Of course they are.

Second step:

Look at the lake shapes, which having the form of zigzag shapes. The zigzag areas, bottom of the left hand side where lake turns sharp right just below the box 5 then down and so on.

Third step,

The Lake measurement starts after 2 mile from the land scape begins.

Fourth step,

If you count the total boxes (Clear Lake shown in fig). The figure boxes can be identified its each column and row numbers. It gives the total area of the lake. The calculation details are furnished herein.

Each box holds <u>4 squared-miles</u> (sq. miles).

CBEST Math Section:
California Basic Educational Skills Test

1. Location column 2 holds row numbers 5 and 6 = 4x2= **8** sq.-miles
2. Location column 3 holds row numbers 5, 6, and 7 = 4x3 boxes= **12** sq.-miles
3. Location column 4 holds row numbers 3,4,5,and 6 = 4x4 boxes = **16** sq.-miles
4. Location column numbers 5,6,7,(3-column) and hold row numbers 5, 6,7, and 8 (4-row) = 3x4x4= **48** sq.-miles

   Total lake area = (8+12+16+48) = 48

CBEST Math Section:
California Basic Educational Skills Test

CBEST MATH WATCH!

QUESTION FOCUSES ON AREA MEASUREMENT

What is the total area of Clear Lake?

6. **Use the diagram below to answer the question that follows.**

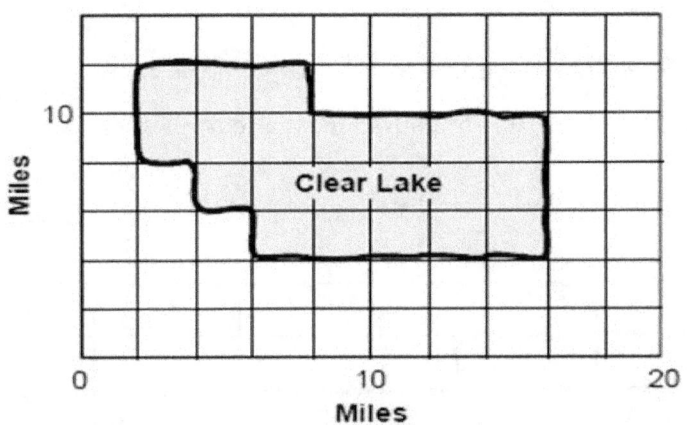

What is the total **AREA** of Clear Lake ?

A. 22 miles

B. 44 miles

C. 48 miles

D. 56 miles

E. 84 miles

CBEST Math Section:
California Basic Educational Skills Test

Alternative approaches to figure out Clear Lake's total area that shown shaded area above.

Step #1: Figure out the each box area. (It already discussed earlier paragraphs).

Step #2: Count the number of boxes.

Step #3: Multiplying number of boxes with the area of the each box area, this measured in squared-miles.

Count the boxes from the diagram provided below:

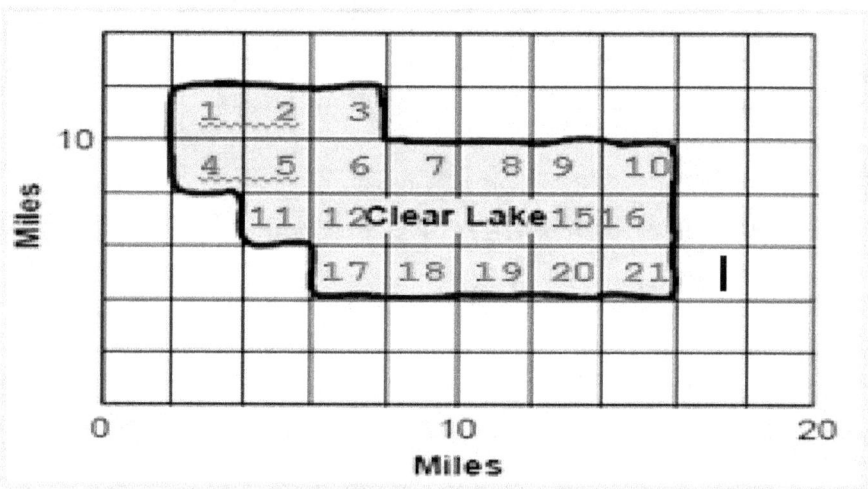

There are 21 boxes in which each holds 4 squared-miles.

So, total area (21 boxes x 4 sq-miles) = 84 squared-miles lake area.

CBEST Math Section:
California Basic Educational Skills Test

CBEST MATH WATCH!

QUESTION #6: Peripheral measurement

6. **Use the diagram below to answer the question that follows.**

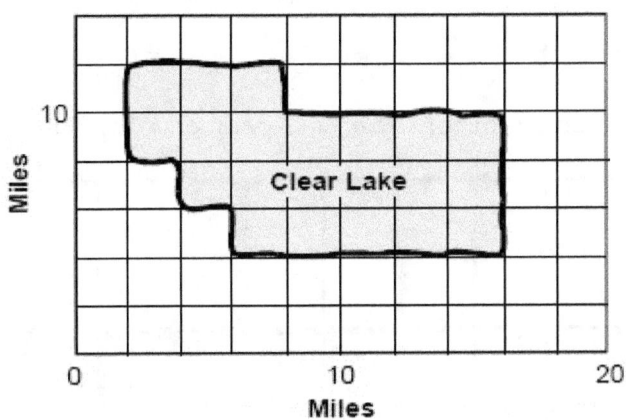

What is the total length of Clear Lake's shoreline?

A. 22 miles

B. 44 miles

C. 48 miles

D. 56 miles

E. 84 miles

This question wanted to know the Clear Lake's perimeter.

The keyword is perimeter that represent the lake's shoreline.

CBEST Math Section:
California Basic Educational Skills Test

To solve the perimeter, it needs count all sides of the diagram. For example, clear Lake Shoreline. Shoreline is the perimeter.

Look at the following diagram shown below, which has the great impact to figure out the perimeter of Clear Lake shoreline.

Measurement of the Clear Lake perimeter diagram

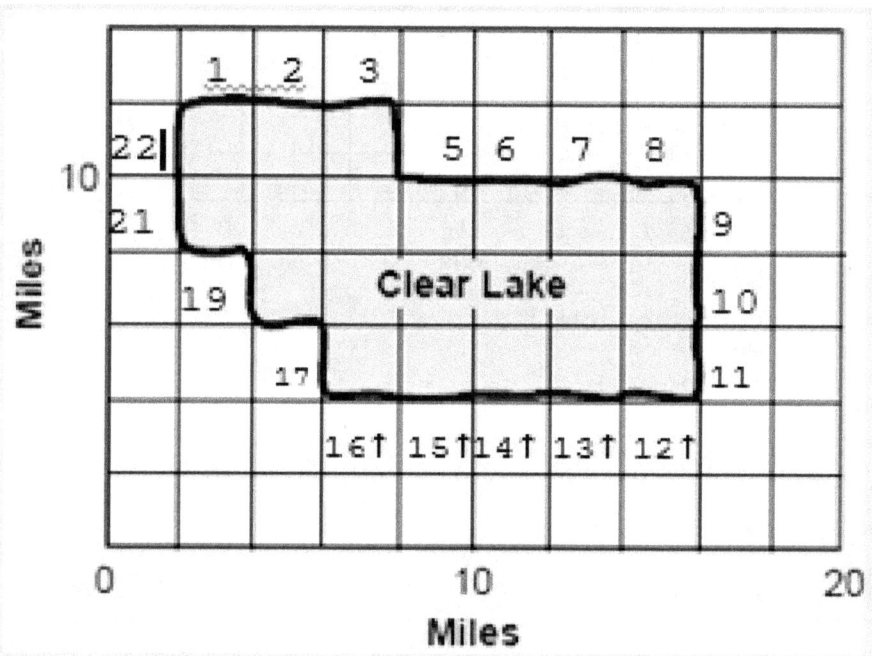

There are 22 counts, which have marked 1 through 22. Each side length is 2 miles.

Therefore, 22x2 = 44 miles perimeter of the shoreline

The choice is B.

CBEST Math Section:
California Basic Educational Skills Test

CBEST MATH TEST WATCH!

QUESTION #4: Length Measurement using feet and inches.

Ref:

http://www.ctcexams.nesinc.com/PDF/CBEST_OP T_Math.pdf

5. Ms. Gutierrez needs to order rope for her gym class of 32 students. Each student will receive a piece of rope that is 5 feet 8 inches long. What is the total length of rope Ms. Gutierrez needs to order for her class?

   A. 106 feet 8 inches

   B. 154 feet 8 inches

   C. 160 feet 8 inches

   D. 181 feet 4 inches

   E. 185 feet 6 inches

Math Analysis:

Ms. Gutierrez's class has 32 students. Each student needs 5 feet 8 inches long rope.

Therefore, 32 x 5'8"

To solve this question, we have to change feet into inches and then change total inches to feet and so on.

CBEST Math Section:
California Basic Educational Skills Test

Change feet into inches then adding 8 inches, which becomes (5 feet x 12) = 60" + 8" = 68"

So, one piece of rope length is 68"

Ms. Gutierrez class holds 32 students.

Total length needed for the class = (32x68) = 2176 inches long rope.

As we know, 12" = 1'

Therefore, $\frac{2176}{12}$ = 181.33 feet

181.33 = 181 feet 4 inches (0.33 = 4")

The answer choice is D.

Basic Probability Principles:

Probability:

What is probability?

In mathematics, probability is chance likely to occur via an event in which measures by the ratio of the favorable cases to the whole number of cases possible.

$$\text{Probability} = \frac{Number\ of\ Favorable\ Outcomes}{Number\ of\ Possible\ Outcomes}$$

For example,

Using a dime-coin toss, what is the probability of tossing outcomes? A dime has one side is head and opposite side is tail. So, one toss results would be either head or tail. The probability is only one side i.e. head or tail.

The possible outcome is:

Probability $= \frac{1}{2}$.

CBEST Math Section:
California Basic Educational Skills Test

Learn from Examples:

Example 1:

What is the probability of tossing <u>heads</u> 4 consecutive times with a quarter-coin?

A $\frac{1}{2}$

B $\frac{1}{3}$

C $\frac{1}{4}$

D $\frac{1}{8}$

E $\frac{1}{16}$

The answer choice is E.

Why choice E is the answer?

Because each tossing event is separately or independently took place. The odds are $\frac{1}{2}$ for each toss, which can be put this way:

Probability = $\frac{1}{2} \times \frac{1}{2} \times \frac{1}{2} \times \frac{1}{2} = \frac{1}{16}$

CBEST Math Section:
California Basic Educational Skills Test

Example 2:

A jar contains 5 dime, 4 nickel, and 3 pennies.

What is the probability of random selecting a penny on the first draw?

A. $\frac{5}{13}$

B. $\frac{4}{13}$

C. $\frac{3}{13}$

D. $\frac{2}{13}$

E. $\frac{7}{13}$

Total coins respectively are dime, nickel, and penny 5, 4, and 3. Total coins are together (5+4+3)=13, which is total number of possible outcomes.

The probability is looking at random selecting a penny. It means a penny will be found among 3 pennies from the jar. So, number of favorable outcomes expected 3/13. The answer is C.

CBEST Math Section:
California Basic Educational Skills Test

Example 4:

What is the probability of spinning a 5 in single spin?

A. 1

B. $\frac{1}{2}$

C. $\frac{1}{5}$

D. $\frac{1}{7}$

E. $\frac{1}{10}$

As we see the wheel it is equally divided into 10 pies. Thus the probability is $\frac{1}{10}$.

Probability of spinning a 5 in single spin may or may not occur, but not guarantee.

CBEST Math Section:
California Basic Educational Skills Test

Predicted Outcomes:

Arithmetic Combinations:

In mathematics, combination is selecting several items out of a larger group of items, where permutations order does not issue. In smallest case, for example, 3 marbles grouping, in such case these 3 colors red, green, and blue can be illustrated grouping of combinations:

i. (Red and green)

ii. (Red and blue)

iii. (Green and blue).

These 3 groups are the total number of possible combination choices.

Learn from Examples:

Example 1:

There are 3 kinds meal and 5 different soft drinks available for lunch at Lily fast foods restaurant. How many ways that 3 kinds of meal and drinks can be served among the consumers?

    A 10
    B. 12
    C. 15
    D. 20
    E. 25

Arithmetic combination is simply multiply the 3 and 5, which becomes 15.

The answer is C.

CBEST Math Section:
California Basic Educational Skills Test

Permutations:

What is permutation?

In mathematics, a number of successive choices (S T A R T) to make calculation and the choice are affected by the previous (S <- T), (T <-A), (A <- R), (R <-T) choice(s), where the order is important part of permutation calculations.

Remember, when the order is not matter, it is not a permutation, but a combination.

For example,

How many ways "S T A R T" can be arranged in a row?

S T A R T consists of 5 letters. In arithmetic, it symbolizes 5!

Permutation is, $5! = 5 \times 4 \times 3 \times 2 \times 1 = 120$

Thus,

5 is the first choices for first letter - S

4 is the first choices for second letter - T

3 is the first choices for third letter - A

2 is the first choices for fourth letter - R

1 is the first choices for fifth letter – T

Permutation Example 2:

Three place in order arrangement permutations.

Statement:

Jose wants to visit 3 places; Universal Studio (U), Magic Mountain (M), and Hollywood (H).

How many choices does John can make?

UMH, UHM, MUH, MHU, HMU, HUM

Jose can decide one of the choices among 6 available options.

CBEST Math Section:
California Basic Educational Skills Test

Algebra

Basic Algebra	190
Algebraic Symbol & Sign	194
Algebraic Order of Operation	198
Algebraic Radial Simplyfing	200
Algebraic Equation Polynomials	203
Factoring Square Bionomials	205
Multiplying Factoring	206
Factoring Polynominals	208
Factoring Polynominals Step 2	210
Algebraic Variable	211
	214
Graphs, Charts, & Tables	216
Graphing	217
Bar Graph Charts	220
Percentage Calculation	226
Line Graph	228
Circle Graphs:	237
Introduction	237
Pie Charts	244
Table Charts:	251
	253
Prediction, Comparision, Computation	253
	254

CBEST Math Section:
California Basic Educational Skills Test

Basic Algebra

Simple equation:

$-5+(-7)= -12$;   Adding -5 and -7 becomes -12

It means $- + - = -$ minus sign precedence. Rule#1

$x+(-y)=x-y$

$+ x$ adding $-y = -$ minus sign also precedence. Rule #2

Explanation of algebraic rules:

Minus – symbol or sign.

- minus - sign or symbol represents negative numbers, for example, -5 is known as negative 5. Its opposite is +5 or 5

- Negative fraction is $-\frac{3}{4}$. It known as negative three-fourths. Its opposite value is $\frac{3}{4}$.

- When symbol appears between 2 value, it is known as subtraction – sign. For example, $5-2 = 3$;

    $-$ opposite is $+$

    $-x$ opposite $+ x$ or x

    for example, $-(x+3)$ is opposite $(x+3)$ or $x+3$

CBEST Math Section:
California Basic Educational Skills Test

- Exponential example

-3^2 opposite 3^2

Algebraic Graphic Representation & Relations:

Scale measurements

X-axis Example,

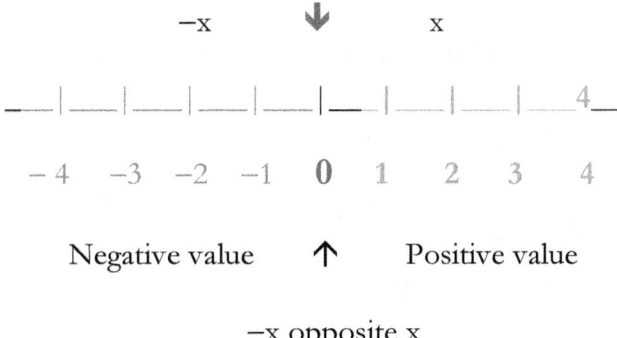

−x opposite x

Absolute Value:

The absolute symbol is a pair of vertical 2 lines | | that surrounding a number, for example, $|-7| = 7$; $|-5| = 5$; $|-3| = 3$.

From three examples, we learned that absolute value never negative. So absolute value is always zero or a positive number.

CBEST Math Section:
California Basic Educational Skills Test

Algebraic Signs and Symbols

Multiplication * or x Sign:

Learn from Examples:

- $6 * (-7) = -42$, it means that if 2 sign are deferent, the product is negative. Where 6 is positive, and -7 negative number.

- $-7(-6) = 42$, it means that if 2 sign are same, the product is positive.

- $(-9) * (-5) = 45$, this equation has 2 same thus, its product is positive value.

Algebraic Division ÷ Sign Usage:

$(-42) ÷ (-6) = 7$, it means that if 2 sign are same the dividing output is positive.

$(-42) ÷ (6) = 7$, it means that if 2 sign are different, dividing output is negative.

CBEST Math Section:
California Basic Educational Skills Test

Exponents:

One exponential value look like this: 3^2

It is 3 power 2.

The base value is 3, and the raise number 2 is the power, which called the exponent. In other word, 3 factors of 2;

$3 * 3 = 9$

Multiplication rule of exponential values:

Rule #1: if 2 sign are same the product is positive, for example"

$3^2 = (-3) * (-3) = 9,$

Rule#2. if 2 sign are same the product is positive, for example:

$3^2 = (3) * (3) = 9$

Multiplication of two values, which contain negative and positive exponential value.

$(-3^2) * (3^2) = (9) * (9) = 81$

Learn from examples:

5^0, it is 5 power 0 (zero). Any number in this example has 1 value except o.

Therefore, $5^0 = 1$

5 power 3, which can be written like this 5^3

$5^3 = 5 * 5 * 5 = 125$

-5 power 3, which can be written like this -5^3

$-5^3 = (-5) *(-5) *(-5) = -125$

Square Root:

What is square root?

The **square root** of a number i.e. a value, when that number multiplied by itself, it gives the number, for example, $7 * 7 = 49$,

Square root symbol is this: $\sqrt{}$

The number 49 square root $\sqrt{49} = \sqrt{(7*7)} = 7$

A squared root of $\sqrt{49} = 7$

Learn from examples:

$\sqrt{25} = \sqrt{5^2} = 5$

$-\sqrt{25} = \sqrt{\{(-5).(-5)\}} = 5$

$-5^2 = \sqrt{\{(-5).(-5)\}} = 5$

$5^2 = \sqrt{(5.5)} = 5$

$\sqrt{36} = \sqrt{(6.6)} = 6$

$-\sqrt{36} = \sqrt{\{(-6).(-6)\}} = 6$

CBEST Math Section:
California Basic Educational Skills Test

The above examples contain −5, 5, and −6, 6, which have square roots value respectively 25 and 36. They contained positive numbers. (−6)(−6)= +36 . (Recall multiplication rules that discussed earlier).

$\sqrt{0}=0$, zero does not have value

a negative value that followed by − negative signs have no square root operation, for example, $\sqrt{-9}$. There are no square roots of negative numbers.

Learn from examples:

$\sqrt{64} = \sqrt{(8*8)} = 8$

$\sqrt{0} = 0$

$\sqrt{-81}$ no functional square roots

Because − 9*(-9) = 81

CBEST Math Section:
California Basic Educational Skills Test

Algebraic Order of Operation:

There are several steps to do algebraic order of operation that stated hereafter:

Step #1

- Parentheses first. What it means is, anything such as inside the parentheses ( ), brackets [ ], curly brackets { }, etc. grouping must be addressed first and solve before proceed next step.

For examples:

$$-6 - 2 [5 - \{ 4 - (1-4) \}]$$

$$-6 - 2 [5 - \{ 4 - (-3) \}]$$

$$-6 - 2 [5 - \{ 4 + 3 \}]$$

$$-6 - 2 [5 - \{ + 7 \}]$$

$$-6 - 2 [5 - 7]$$

$$-6 - 2 [-2]$$

$$-6 + 4$$

$$-2$$

CBEST Math Section:
California Basic Educational Skills Test

Step#2

- Solve all square roots and exponents left to right order.

For examples:

    a. $\sqrt{36} = \sqrt{6} \cdot \sqrt{6} = 6$

    b. $\sqrt{(49 + 36)} = \sqrt{\{(7.7)+(6.6)\}} = 7+6 = 13$

Step#3

- Solve multiplication and division from left to right order

Examples:

$$-7 + 4 + (7 - 8)$$

$$= (-7+4) + (7-8) \quad \text{solve left}$$

$$= (-3) + (7-8) \quad \text{solve next right}$$

$$= (-3) + (-1) \quad (+ * -) = - \text{ minus precedence}$$

$$= -3 + (-1)$$

$$= -3 - 1$$

$$= -4$$

CBEST Math Section:
California Basic Educational Skills Test

Step#4

- Finally solve addition and subtraction from left to right.

Examples:

$$7 + 4(7-8) = 7+4(-1) = 7-3 = 4$$

RADICAL SIMPLYFING:

What is radical?

Radical expresses about a square root, cube root, etc.

The symbol it usage is √ (root).

Radical has few components such as index, radical symbol √ and radicand.

When a radical contain a square root, it should be simplified but number 1.

All radical must be simplified, for example,

$$\sqrt[3]{27 \cdot x^3 y^3}$$

$$\sqrt[3]{3^3 \cdot x^3 y^3} = 3bc$$

CBEST Math Section:
California Basic Educational Skills Test

Learn from examples:

- $\sqrt{(x^2 \cdot y^2)} = \sqrt{(x \cdot x)} \sqrt{(y \cdot y)}$

  $= xy$ (simplified)

- $\sqrt{1000} = \sqrt{100 \cdot 10} = 10\sqrt{10}$ (simplified)

- $\sqrt{500} = \sqrt{100 \cdot 5} = 10\sqrt{5}$ (simplified)

- $\sqrt{72} = \sqrt{36 \cdot 2} = 6\sqrt{2}$ (simplified)

- $\sqrt[3]{27} = \sqrt[3]{(3 \cdot 3 \cdot 3)}$

  $= \sqrt[3]{3^3} = 3$ (simplified)

- $3\sqrt{75} = 3\sqrt{(25 \cdot 3)}$

  $= 3\sqrt{25}\sqrt{3}$

  $= 3\sqrt{5^2}\sqrt{3}$

  $= 3 \cdot 5\sqrt{3}$

  $= 15\sqrt{3}$ (simplified)

- $2/3\sqrt{27} = 2/3\sqrt{9 \cdot 3}$

  $= 2/3\sqrt{(3^2 \cdot 3)}$

  $= 2/3(3) \cdot \sqrt{3}$

  $= 2\sqrt{3}$ (simplified)

- $2/3\sqrt{24} = 2/3\sqrt{4.6}$

$= 2/3\sqrt{2^2}.\sqrt{6}$

$= 2/3(2)\sqrt{6}$

$= 4/3\sqrt{}$

Algebraic Simplification:

It's dealing with distribution property, which involves combining properties by eliminating parentheses.

For example, $x(y+z) = xy + xz$

Learn from examples:

- $5(5x - 6) = 25x - 6$
- $3(x - 5) + 4(x + 6)$

    $= 3x - 5x + 4x + 24$

    $= (3x + 4x) - 5x + 24$

    $= 7x - 5x + 24$

    $= 2x + 24$ (simplification outcomes)

CBEST Math Section:
California Basic Educational Skills Test

Algebraic Equation:

There are few concepts in order to solve algebraic equations.

- Adding and/or subtracting equation that should be equally balanced both sides of the equations.

For examples:

$$5x - 1 = 6$$

or $5x = 6 + 1$ (Adding)

or $5x = 7$

or $x = 7/5 = 1\ (2/5)$ (dividing)

- Multiplying and/or dividing that must be balanced both sides of the equations that make sense.

For examples:

- $x/5 = (x/4) - (½)$

or $(x/5) - (x/4) = - (1/4)$

or $20\ \{(x/5) - (x/4)\} = \mathbf{20}\{- (1/4)\}$

or $20\ (x/5) - 20(x/4) = \mathbf{20}\{- (1/4)\}$

CBEST Math Section:
California Basic Educational Skills Test

or $4x - 5x = -5$

or $-x = -5$

or $x = 5$

Learn from examples:

- $(x+2)3 + (x+1)/2 = 7/6$

  or $\{(x+2)/3\} + \{(x+1)/2\} = 7/6$ ((multiply 6 by both sides)

  or $6[\{(x+2)\}/3] + 6\{(x+1)\}/2] = 6(7/6)$

  or $2\{(x+2)\} + 3\{(x+1)\} = 7$

  or $2(x+2) + 3(x+1) = 7$

  or $2x + 4 + 3x + 3 = 7$

  or $5x + 7 = 7$

  or $5x = 7 - 7$

  or $5x = 0$

  or $x = 0$

Polynomials:

Addition & subtraction

Polynomials dealing with addition and subtraction, which are combining similar terms.

Examples

$$4x - 2x - 4$$
$$\underline{3x - x - 2}$$
$$7x - 3x - 6 = 4x - 6$$

Multiplication Algebraic Binomials:

$(3x + 2)(4x - 5)$

$= 3x(4x - 5) + 2(4x - 5)$

$= 12x^2 - 15x + 8x - 10$

$= 12x^2 - 7x - 15$

The above solution is used common distribution property, which known as FOIL method.

F O I L stands for **F**ast, **O**uter side, **I**nside, and **L**ast.

Factoring Square Binomials:

1. Square Binomials is the square of the first term, plus the square of the last term, . plus twice the product of the terms. Formula#1 of factoring that expresses as follows:

    Formula #1

    - $(x + y)^2 = x^2 + y^2 + 2xy$

    Example,

    $(x + 3)^2 = x^2 + 3^2 + 2.3.x$

    or $x^2 + 9 + 6x$

    or $x^2 + 6x + 9$

2. Square Binomials is the square of the first term, plus the square of the last term, minus twice the product of the terms.

    Formula#2

    - $(x - y)^2 = x^2 + y^2 - 2xy$

Example,

$$(x - 3)^2 = x^2 + 3^2 - 2.3.x$$

$$\text{or} = x^2 + 9 - 6x$$

Factoring:

Algebraic factoring is a process of changing a polynomial into product of other polynomials.

Common Factor:

When a polynomial's all the terms contain a common greater factor, use the distributive property to remove the largest/greater common factor.

For example, factoring equation this:

$4x - 16$.

In this equation has largest common factor is 4, which uses distributive property.

$$4x - 16 = 4(x - 4).$$

Learn from examples:

- $25x^3 - 20x^2$

or $5x^2 (5x - 4)$ 5

$5x^2$ is the greatest common factor that separated from the given, $25x^3 - 20x^2$, polynomials.

Multiplying Binomials:

Multiplying a sum (addition) and a difference (subtraction) outcomes.

The product of the sum $(x+y)$ and the difference $(x-y)$ of two terms is the square of the first term (x^2) minus $(-y^2)$ the square of the second. It illustrates like this:

$$(x + y)(x - y) = x^2 - y^2$$

CBEST Math Section:
California Basic Educational Skills Test

For examples:

1. $(x + 5)(x - 5)$

    $= x^2 - 5^2$

    $= x^2 - 25$ (simplifying).

2. $(x^2 - 7)(x^2 + 7)$

    $= x^2(x^2) - 7(7)$

    $= x^4 + 49$

Factoring Polynomials type 1:

$x^2 + bx + c$

To solve this type of factoring follow the steps given bellow.

Step#1:

First coefficient (x^2) polynomial term is always considered 1.

Step#2

The last term c, whose sum (if +ve sign) or difference (if –ve sign) is equal to the coefficient of the middle term.

Step#3

These numbers are the second terms of the two binomials

Let looks how it works!

An example is: $x^2 - x - 6$

$x^2 - x - 6$, its first coefficient is x^2 that consider 1 and last coefficient is – 6, which has –ve sign. Therefore, the middle terms would be -3 . +2 = - 6.

Let's organize them into type#1 formula.

$$x^2 - x - 6$$

$$= x^2 - 3x + 2x - 6 \text{ (first term is 1 and last term is - 6)}$$

$$= x(x - 3) + 2(x - 3)$$

$$= (x - 3)(x + 2)$$

Factoring Polynomials type 2.

$ax^2 + bx + c$

To solve this type of factoring follow the steps given bellow.

Step#1:

First coefficient a (a from ax^2) polynomial term is not 0 or 1. It can't $a \neq 0$ and $a \neq 1$

Step#2

Multiply the first term (a) and the last term (c), whose sum (if +ve sign) or difference (if –ve sign) what may be is equal to the coefficient of the middle term.

Step#3

These numbers are the second terms of the two binomials

CBEST Math Section:
California Basic Educational Skills Test

Let looks how it works!

An example is: $2x^2 - x - 15$

$2x^2 - x - 15$, its first coefficient is a is 2 and last coefficient is -15, which has –ve sign. Therefore, the middle terms would be $-6 \cdot +5 = -1$.

Let organize them into type#1 formula.

$2x^2 - x - 15$

$= 2x^2 - 6x + 5x - 15$ (first term is 2 and last term is - 15)

$= 2x(x - 3) + 5(x - 3)$

$= (x - 3)(2x + 5)$

Learn from examples:

Factoring equation 1

- $2x^2 + 13x + 18$

   (first term 2 and last term 18, which product is + 36. The middle give terms 12x that to be equal to the coefficient of the middle terms. It therefore, $(+9) * (+4) = +36$

$2x^2 + 13x + 18$

$= 2x^2 + 4x + 9x + 18$

$= 2x(x+2) + 9(x+2)$

$= (x+2)(2x+9)$

Factoring Equation 2:

- $2x^2 + 12x + 18$

(first term 2 and last term 18, which product is + 36. The middle give terms 12x that to be equal to the coefficient of the middle terms.

It therefore, $(+6) * (+6) = +36$

$2x^2 + 12x + 18$

$= 2x^2 + 6x + 6x + 18$

$= 2x(x+3) + 6)x + 3)$

$= (x+3)(2x+6)$

Algebraic Variable

Variable represent unknown thing with an equation, for example, $x+3 = 5$.

x is a variable of this equation that is unknown at this point until we solve it.

Let's try, $x+3 = 5$

$x = 5 - 3$ (change the 3 opposite of the = sign becomes –ve value of 3 or -3.

$x = 5 - 3 = 2$

the variable in this scenario is 2.

Learn from examples:

If x is positive integer in the equation $3x=y$, then what should be Y? $3x = y$

Let says, $x = 2$

$3x = y$ or $3*2 = y$

or $6 = y$

or $y = 6$

CBEST Math Section:
California Basic Educational Skills Test

This Page is Internationally Blank

CBEST Math Section:
California Basic Educational Skills Test

Graphs, Charts, and Tables

Introduction

Math (graphs, charts, and tables) section, the test takers must be able to understand and demonstrate their abilities to the following areas; graphs, charts, and tables.

GRAPHS

Charts display many ways.

The CBEST Math focuses on mainly 3 graphical charts.

They are

- Bar graphs
- Line graphs
- Circle graphs

Graphing

Most common graphical method used in algebra is known as coordinate (rectangular representation) system.

```
                    y ↑
           II              I
      -x ←  x-axis         →
      ─────────────┼─────────────
           III    -y ↓    IV
                 y- axis
```

The graph has horizontal line that represents x-axis, and vertical line that represent y-axis. These two lines separate the plane into four quadrants numbered. I, II, III, and IV.

This rectangular coordinating system graphs helps to identify the points on the graphs with ordered pairs of real numbers.

An order of pair consist of two number by a comma, set within parentheses like this (x, y) or/and (-x, -y). The line x-axis measure horizontal , and y-axis measure vertical line.

CBEST Math Section:
California Basic Educational Skills Test

So, if the pair is (2,1) it means x point is located 2 points positive direction of the graph and y is located 1point positive direction where they met together, the point test-taker is looking for to satisfy the response.

For example, $2x + 5y - 9 = 0$

Where is the source of the pairs of this equation with its two variable where pair met point (2,1)?

$2x + 5y - 9 = 0$

or $2x + 5y = 9$

or $2x + 5y = 9$ (replace x and y value from (2,1)

or $2.2 + 1.5 = 9$

or $4 + 5 = 9$

or $9 = 9$, the result is true

When it evaluated (x, y) respectively (2,1), the equation's both sides test proved true.

Learn from examples:

Evaluate the $2x + 5y - 9 = 0$ when $x = 6$

$2x + 3y - 9 = 0$

or $2.6 + 3y - 9 = 0$

CBEST Math Section:
California Basic Educational Skills Test

or $12 + 3y = 9$

or $3y = -12 + 9$

or $3y = -3$

or $y = -3/3 = -1$

It prove that pair (x, y) respectively (7, -1) when x replaced by 4.

Test the following equation when x is -1

$2x + 3y - 9 = 0$

or $2.(-1) + 3y - 9 = 0$

or $-2 + 3y = 9$

or $3y = 9 + 2$

or $3y = 11$

or $y = 11/3$

Thus, (x, y) represented (-1, 11/3)

CBEST Watch! $3x - 2y = 0$, if (-3, 2)

$3x - 2y = 0$

or $(-3*3) - (2*2 = 0$

or $-9 - 4 = 0$

or $-13 = 0$, not true.

CBEST Math Section:
California Basic Educational Skills Test

- Find missing pair if (4, ?), (7, ?), (-3, ?)

    $5 = x + 4y$

    or $5 = 4 + 4y$

    or $4 + 4y = 5$

    or $4y = 5 - 4$

    or $4y = 1$

    or $y = 1/4$

    (x, y) respectively represents (4, ¼)

BAR GRAPHIC CHARTS:

Introduction

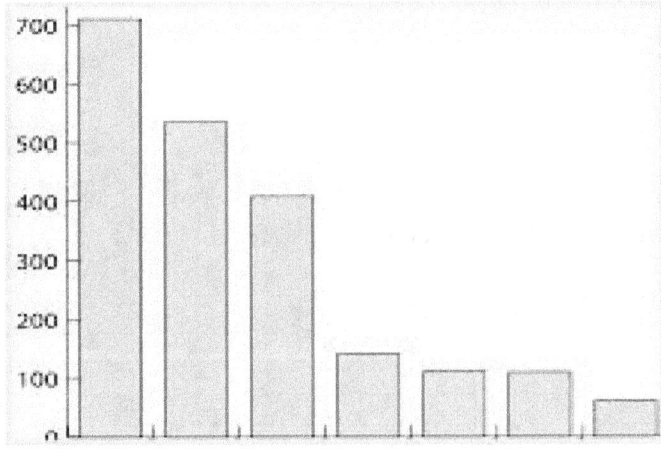

Vertical View Bar Graphs

CBEST Math Section:
California Basic Educational Skills Test

Bar chart has two axes.

1. Vertical bar. It represents Y axis.

2. Horizontal bar. It represents X axis.

These two bars represent two (vertical & horizontal) values that are interdependent in which combined a valuable result. It uses mathematical statistical results.

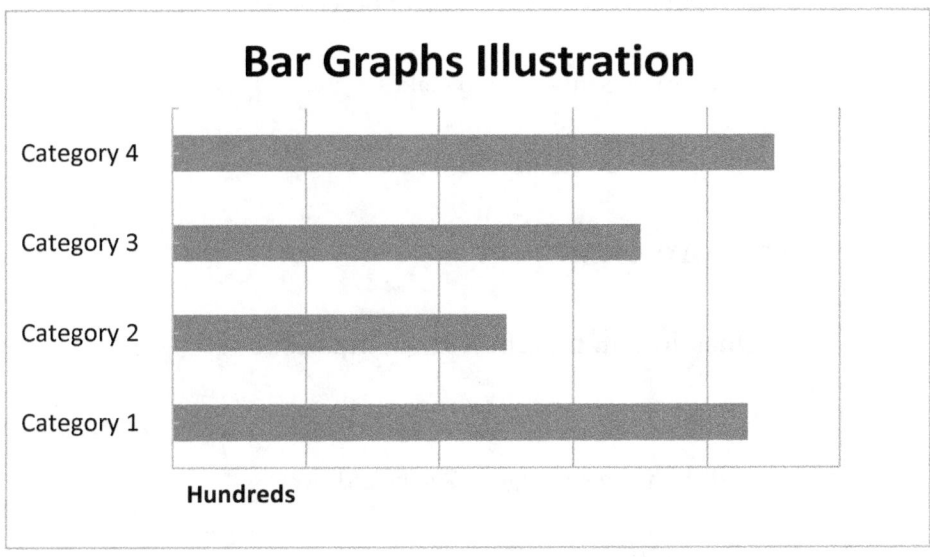

Horizontal View Bar graphs

Analysis of bar graphs:

The chart bars given above is an example of a bar graphs that contain separate bar rows and columns. Each bar and column have its own information, for example, vertical or Y-axis contains information category 1 through 4. Category is the title of the vertical line information can represent product information like

CBEST Math Section:
California Basic Educational Skills Test

Other side is horizontal line or X-axis that starts one hundred through 5 hundreds.

Let's say,

Category 1 contains 425 hamburgers

Category 2 contains 250 salad disks

Category 3 contains 350 Cakes

Category 3 contains 450 Cookies

Learn from example!

Question Statement #01: Comparison

Hamburger category 1 has approximately how many more sold than does category 2 salads disk?

    A. 150

    B. 175

    C. 200

    D. 400

    E. 450

The best answer is B

How's it?

CBEST Math Section:
California Basic Educational Skills Test

It needs graphic bar length reading ability. Look at the horizontal line value that denoted number value from left side to right. It starts 0 and ends 500. Each line interval is 100. Compare two categories that asking question related answer i.e. category 1 and 2.

Test takers must understand how graphic interval is divided and its scale. Each line holds value (number in hundreds or/and thousands etc.

If test takers able to read correctly graphical representation of horizontal and vertical length (scale that represent number or some other unit) then it is very simple to overcomes this type of question.

The answer analysis it provided hereafter:

Let's move for example how it works!

Category 1 is hamburger item sold	425
Category 2 is hamburger item sold	250
Substrate value is (425 – 250)	**175**

CBEST Math Section:
California Basic Educational Skills Test

Typical CBEST graphical math puzzle illustration

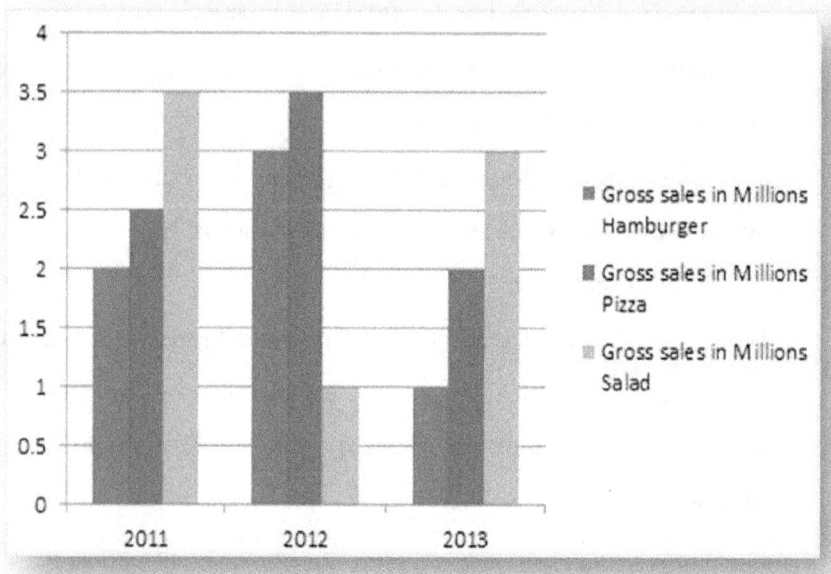

1 million = 1,000,000

1.5 million = 1,500,000

0.5 Million = 500,000

Test takers have to have ability to read the graphic bars' length or height. It focuses on *approximate* reading ability

CBEST Math Section:
California Basic Educational Skills Test

Typical CBEST question:

Question #01:

The 2011 gross sales receipts of Hamburger, Pizza, and Salad exceeded that same first food items in the year 2013?

    A. 0.4 million

    B. 1.9 million

    C. 8.0 million

    D. 6.0 million

    E. 14.0 million

The best approximate answer is **B**.

It is the approximate answer. Because, the year 2011 gross is (2.0+2.5+3.5) = 8.0 million.

The year 2013 gross is (1.0+2.0+3.0) = 6.0

Therefore, the difference of those two years gross sales is (8.0–6.0) = 2.0 million.

CBEST Math Section:
California Basic Educational Skills Test

Question #02:

From 2011 to 2013, the percentage increase/decrease in receipts for those 3 first food items reduced the percent decrease by approximate how much?

A. 76%

B. -43%

C. 33%

D. -24%

E. 10%

The best approximate answer is D.

Percentage Calculation Formula:

$$Percentage\ Change = \frac{Amount\ of\ the\ change}{The\ starting\ amount\ (followed\ the\ word\ from)}$$

Gross receipt sales in the year 2011 = 8.0 million

Gross receipt sales in the year 2013 = 6.00 million

Amount of the change decline = 2.0 million

Decline considered negative transaction; therefore, it can denoted -2.0 million.

CBEST Math Section:
California Basic Educational Skills Test

The starting or from amount is $= \dfrac{-2.0}{8.0} = -25$

or -25%

The answer is negative sales of – 25% that is nearest or approximate decrease of – 24%.

Question #03:

The year 2012 Salad sales declined may be attributed to one of the following statements.

Which statement is the best response?

A. A decrease in the popularity of hamburgers sales
B. A decrease in the popularity of pizza sales
C. A predictable slump attributable to the increase of pesticide uses.
D. Cannot be determine from the information is provided
E. An increase in the popularity of Burger king gone private.

The best answer is **D**.

The graph does not provided information that suitable match with the statement given except D.

All choices are irrelevant except D, which beyond the information given.

LINE GRAPHICAL CHARTS

Introduction

Line graphs have number of points that depicts below, which converts data into points on a grid. These points are connected to a relation among various data items, for example, data like dates and times etc.

Line chart has two components

1. Vertical line 2. Horizontal line

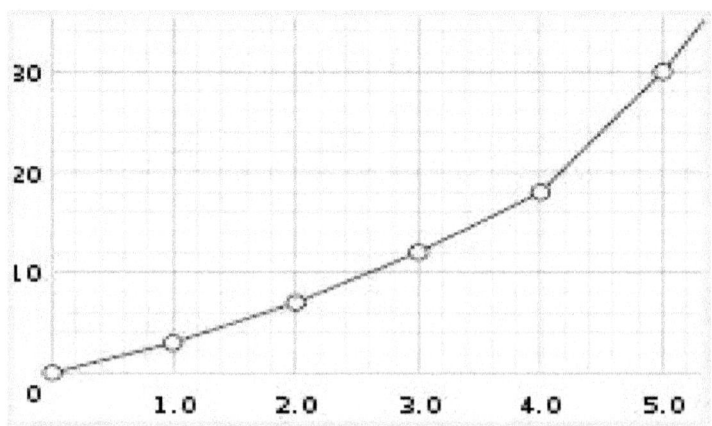

Typical Line Graphs

Let's know some information about the above graphs. Its vertical line starts 0 and ends 30 and horizontal line starts 0 and ends 5.0

The points can be upward slope i.e. increase or downward i.e. decrease. The sharper the slope upward means greater the decrease and sharper downward means greater decrease. The change in data points indicates line graphs trend over a period of time.

CBEST Question Samples:

Use the graph below to answer the two questions that follow:

U.S. Population in the United States from 1900 to 2013

CBEST Math Section:
California Basic Educational Skills Test

Question 1: Ability to read the graph.

In which of the following years were there about 141.75 million U.S. populations?

A. 1915

B. 1935

C. 1945

D. 1965

E. 1995

The best answer is **C**.

Line Graphs:

Line graph reading ability is very simple. The line starts year 1900. It's the starting point i.e. left side to right, which interval (1900 to 1910) = 10 years. It is also known as X-axis or horizontal-axis.

The second reading issue is, vertical reading or Y-axis. The graph shows the number of U.S. population, which move upward and pointing a mark every 10 years interval except afterward 2010, which ended year 2013.

CBEST Math Section:
California Basic Educational Skills Test

What information it carries?

The information it carries are both year and population census every 10 years. So, it starts year 1900 when U.S. populations were 76,212,168 (76.21 million approximate). The graph information ended year 2013. However, the population of year 2013 is not census so far. It estimated U.S. populations were roughly 316.367 million.

Question 2: Ability to read the graph.

During year 1900 to 2010, which of the following time periods was reported slow populations growth?

A. 1910 to 1920

B. 1930 to 1940

C. 1950 to 1960

D. 1980 to 1990

E. 2000 to 2010

The best answer is B.

CBEST Math Section:
California Basic Educational Skills Test

To find the right answer, read carefully graphical lines that presented along with the vertical and horizontal data points first.

The best answer is B. Because, look first at the graph point 1930 and 1940 (first choice).Its line slope is from 1990 to 1940, which almost looks straight line. Another indication may consider is population growth trends. It can be obtained from subtracting methods. However, graphic reading ability nearest or approximation is the quick way to get right answer.

American College Testing (ACT) college readiness assessment is a standardized test for recently high school passed student. They need ACT score for college admission. The following data are furnished herein from 1997 to 2011.

The column "percentage (%) of students who achieved a 36" shows students' score in % from year 1997 to 2011, which drawn line graph hereafter.

Read the table's data that are specially highlighted. These three columns data will be seen converted into line graph. The CBEST graphic math uses such data to its math test.

CBEST Math Section:
California Basic Educational Skills Test

Year	Number of students who achieved a 36	Number of students overall	% of students who achieved a 36
2011	705	1,623,112	0.04337
2010	588	1,568,835	0.03748
2009	638	1,480,469	0.04309
2008	428	1,421,941	0.03010
2007	314	1,300,599	0.02414
2006	216	1,206,455	0.01790
2005	193	1,186,251	0.01627
2004	224	1,171,460	0.01912
2003	195	1,175,059	0.01659
2002	134	1,116,082	0.01201
2001	89	1,069,772	0.00832
2000	131	1,065,138	0.01230
1999	85	1,019,053	0.00834
1998	71	995,039	0.00714
1997	74	959,301	0.00771

Sources: http://en.wikipedia.org/wiki/ACT_(test)

The line graph shown next page, which comes from the above table, which is ACT test from 1997 through 2011.

The vertical line represents "percent of students"

The horizontal line represents "% of student who have achieved a 36" yearly ACT test conducted.

CBEST Math Section:
California Basic Educational Skills Test

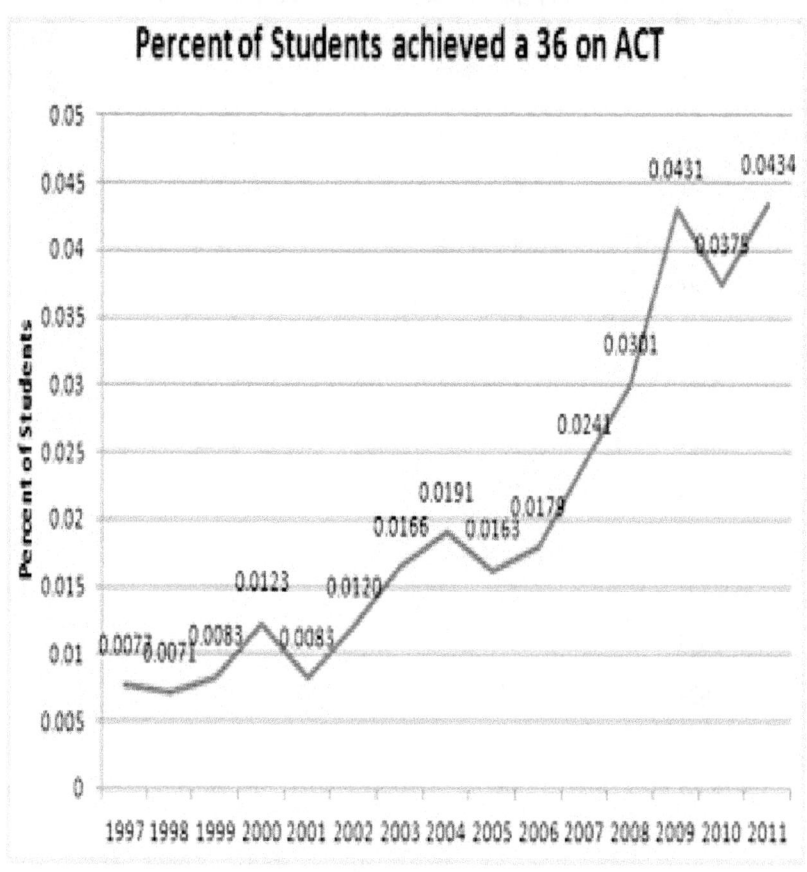

Use the graph above to answer the questions that follow.

CBEST Math Section:
California Basic Educational Skills Test

Question 1:

Between which three years was the greatest rise in average test scores of ACT?

      A. 2000 to 2003

      B. 2003 to 2005

      C. 2005 to 2007

      D. 2007 to 2009

      E. 2009 to 2011

The choice is D. locate the slope of the line graph that moves upward i.e. slope steepest.

Question 2:

In which one of the following years was the highest number of student achievement declined of ACT test?

      A. 1997

      B. 1998

      C. 2001

      D. 2005

      E. 2010

Choice is C.

CBEST Math Section:
California Basic Educational Skills Test

The year 2001 number of student achievement 89 and previous year 2000 was 131. It was the steepest slope on the chart.

Question 3

According to the graph,. which response is correctly depicted.

Which years were the highest numbers of student achievement accomplishment took place of ACT test?

A 1998 - 1999

B 2006 - 2007

C 2007 - 2008

D 2008 - 2009

E 2010 - 2011.

Choice D is correct.

The year 2008 student achievement was 428, which rises 638 in the year 2009. The difference was (638-428) = 210 was the highest number of students' success and highest reading among others.

CBEST Math Section:
California Basic Educational Skills Test

Circle Graphs

Introduction

A circle graph consists of numbers of pie slices. Therefore it is known as Pie charts. The 360 degree circle area is considered 100 percent (100%) area. A circle area can be sub-divided any number of a circle area, which measures as percentage (%) symbol. For example, a pie is one part of the circle that can be 1% area, 2%, 5% and up to 99% area within the 100% circle area. The largest pie or slice is higher percentage like 1% or 99% or 99.99% that way we measure pie or slice's value as percentage area.

The table data found from preliminary results of the election for the European Parliament in 2004.

Group	Seats	Percent (%)	Central angle (°)
EUL	39	5.3	19.2
PES	200	27.3	98.4
EFA	42	5.7	20.7
EDD	15	2.0	7.4
ELDR	67	9.2	33.0
EPP	276	37.7	135.7
UEN	27	3.7	13.3
Other	66	9.0	32.5
Total	732	99.9*	360.2*

CBEST Math Section:
California Basic Educational Skills Test

Learning exercise notes:

The table is presented here for learning purpose. It helps learner how table data are converted into pie charts. The CBEST pie charts questions, the table will not be presented but pie charts. Therefore, do not use this table data when take pie charts multiple choice questions that given test exercise at end of this learning exercise.

About the Table Data:

The table heading are made up of the following:

Group: EUL, PES, EFA, EDD, ELDR, EPP, UEN and other.

Seats: Each group has got the seat like EUL got 39 seats

Percentage (%) each group got that converted into percentage. For example, EUL got 39 out of 732 seats, which comes 5.2% when converted it into percentage by dividing this way: Percent = 39/732.=0.053/100 = 5.3%

The EU total numbers are 732 that made EU parliament.

A circle is fixed 360°. However, the table data shown angle more than 360° because it rounding all central angle columns' data totals 360.2. So ignore angle 360°.

CBEST Math Section:
California Basic Educational Skills Test

The pie charts shown below come from the table data conversion by using spreadsheet application like MS Excel. It is easy to converting pie charts if you use MS Excel database application.

Reference resources:

http://en.wikipedia.org/wiki/File:Pie_chart_EP_election_2004.svg

Typical circle graphs that represents pie charts

This pie graphs represent European Parliament election 2004 results.

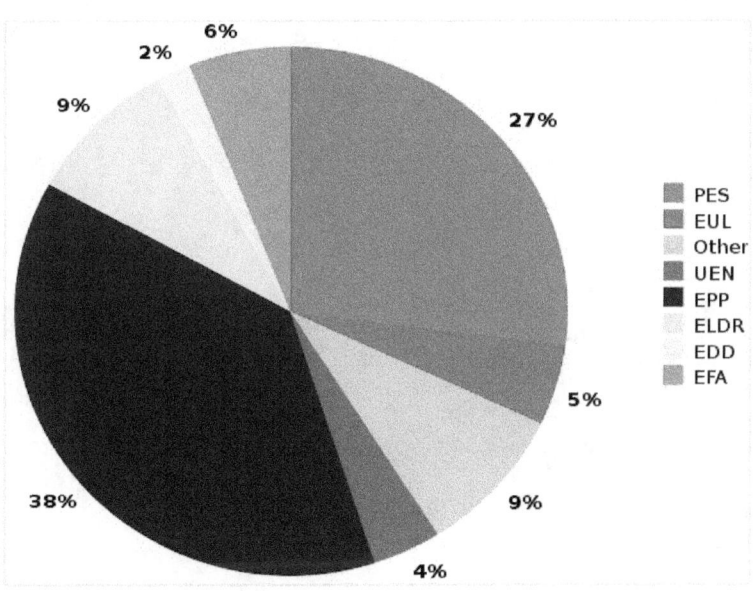

CBEST Math Section:
California Basic Educational Skills Test

The CBEST Sample test questions:

Use the pie graph above to answer the following three questions that follow.

Leaning objective:

Interpreting graphs contents

Question #01:

Which pie represents 9/100 of seats?

A PES

B EUL

C UEN

D EPP

E Other

The pie graphs above given each group's seat numbers along with the percent of seats in parliament. 9 parts or 9% of a circle (a circle is 100%) or it's also 0.09. It represented 9%. It is 'Other'. The correct answer is E.

CBEST Math Section:
California Basic Educational Skills Test

Leaning objective:

Convert a percent into a number

Question #02:

How many seats EPP would represent in the EU Parliament?

A. 200

B. 67

C. 66

D. 276

E. 42

To get this type of question's answer, you must read the graph carefully first. As you've seen that the pie graphs representing percentage of the seats of each group. In this case EPP got 38% that needs converted into seats by dividing it with the total number of seats i.e. 732.

It is a simple math, for example, 100% seat is 732.

EPP got 38% seats that mean $\frac{38}{100}$ part of a circle (a circle is 100%). So, 100% is equal to 732 seats.

Therefore, EPP = $\frac{732 \times 38}{100}$ = 276.

Answer is D.

CBEST Math Section:
California Basic Educational Skills Test

Leaning objective:

Knowledge of Ratio

Question #03:

How do you express the ratio between the EPP seats and the EU parliamentary seats?

A 732:38

B 732:276

C 2.65:1

D 1:2.66

E None of the above

Since mathematical ration is a relationship between two numbers of the same kind. It gives ratio between the same two (732:2760) numbers; however, it is not the answer. Why it is not the answer you've asked? It should be lowest arithmetic values that come from lowest divided points 732/276 = 2.65 i.e. the ratio between EPP and EU parliament seat expressed as 1:2.65 (expressed as fraction is ok). Thus, it is best answer C.

CBEST Math Section:
California Basic Educational Skills Test

Another simple example of ration is,

    4 red marble

    6 blue marble

    2 yellow marble

What are their ratios?

Ratio expresses like this: 4:6:2.

They all can be divided by 2; therefore, their result would be respectively 2:3:1. At this point you can't go further lowest values thus it's their ratio.

Learning Objectives:

Gross sales strategy that uses multiple graphical data

There are four first-food restaurants' sales data given in this exercise. They are

    1. King
    2. CFC
    3. ABC
    4. XYZ

CBEST Math Section:
California Basic Educational Skills Test

All 3 quarters reports shown below are 1st, 2nd and 3rd along with gross sales receipts percentage sales report.

Read all 3 circle graphs' pie data carefully.

Use the graphs below to answer the following question that follow.

Circle Graphs or Pie Charts

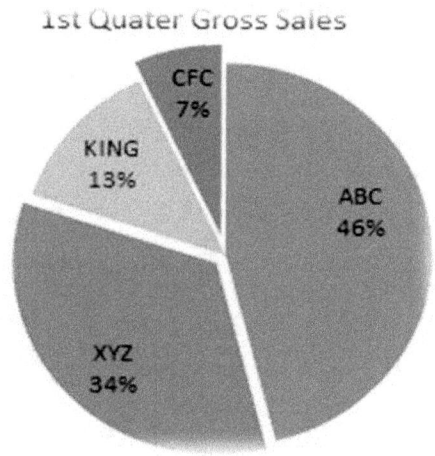

1st Quater Gross Sales
- CFC 7%
- KING 13%
- ABC 46%
- XYZ 34%

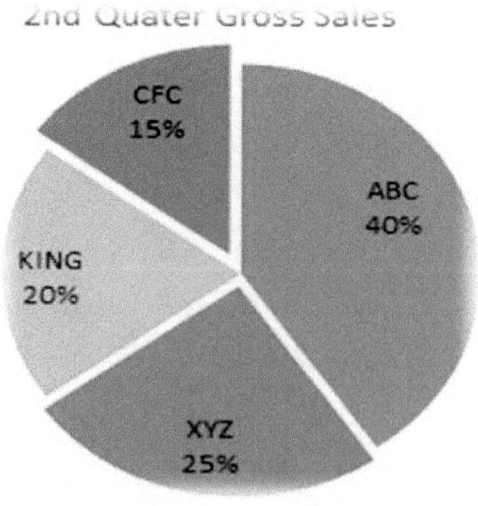

2nd Quater Gross Sales
- CFC 15%
- ABC 40%
- KING 20%
- XYZ 25%

CBEST Math Section:
California Basic Educational Skills Test

3ᴿᴰ Quarter Gross Sales

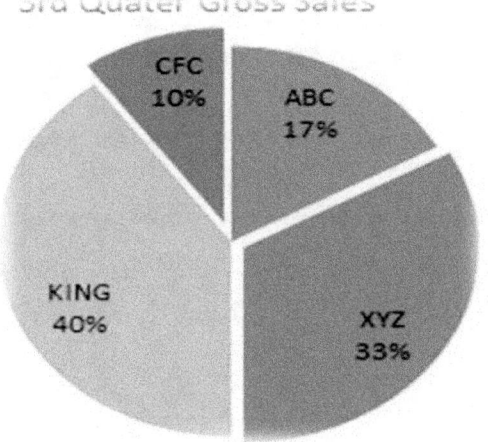

1ˢᵗ, 2ⁿᵈ, and 3ʳᵈ Quarter Gross Sales Receipts in percentage

Combined sales revenue receipts of all four restaurants

1ˢᵗ Quarter Gross sales receipts: 7,400,000

2ⁿᵈ Quarter Gross sales receipts: 8,650,000

3ʳᵈ Quarter Gross sales receipts: 9,100,000

CBEST Math Section:
California Basic Educational Skills Test

Learning objectives:

Vocabs: approximate and rounded off

Response the question bellow based graphical data presented earlier.

The gross sales revenue for 1st quarter is approximately what percentage of the gross receipts for all three quarters?

A 30

B 40

C 45

D 50

E 60

Learn from Step by Step Approach:

Step 1:

Find out the gross sales revenue of all 3 quarters sales amount. Adding up all three quarters revenue like this:

1st Quarter Gross sales receipts: 7,400,000

2nd Quarter Gross sales receipts: 8,650,000

3rd Quarter Gross sales receipts: 9,100,000.

3 Quarters Gross Sales: 25,150,000

CBEST Math Section:
California Basic Educational Skills Test

Write it simplify way that saves time when you take the test because reading/calculating is the critical factors when taking a test. So, write mathematical figure short, for example, 25.15 million instead of 25,150,000.

Step 2:

Gross revenue sale for 1st quarter is 7,400,000 or 7.4 million

Step 3:

$$\text{Percentage} = \frac{7.4}{25.15} = 0.2942 = 29\%$$

The best nearest choice is 30% i.e. A is the right answer.

Learning Objectives:

Not literary meaning, knowledge development of mathematical terminology meaning of; exceeding, average, mean, quarter, percentage etc.

Question #02:

All 3 quarters, average percentage of gross receipts for **ABC restaurant** exceeded the average percentage of gross revenue receipts for **XYZ restaurant** by approximately how much?

A. 50%

CBEST Math Section:
California Basic Educational Skills Test

B. 25%

C. 12.5%

D. 7.75%

E. 3.5%

To calculate the average percentage, it needs adding up all ABC 3 quarters receipts data and then divides by 3. Look at following that shown ABC 3 quarter total 103 and average 34.33. And XYZ 3 quarter total 92 and its average 30.67

	ABC	XYZ
1st Q	46.00	34.00
2nd Q	40	25
3rd Q	17	33
Total	103.00	92.00
Divided by 3Q		
Average	34.33333	30.66667
	or 34.33%	30.67%

ABC exceeds XYZ by 34.33 − 30.67 = 3.67%

The answer choice is **E**.

CBEST Math Section:
California Basic Educational Skills Test

Learning Objective:

Math vocabularies meaning of words: many, much, most, earn, and precise

Question #03:

The gross revenue receipts earn by CFC restaurant in 3rd quarter amount to precisely how much money?

A. 910,000

B. 9,100,000

C. 7,400,000

D. 8,500.000

E. 6,500,000

CBEST Math Section:
California Basic Educational Skills Test

Recall the pie graphs information!

3rd Quarter Gross sales receipts: $9,100,000 is given.

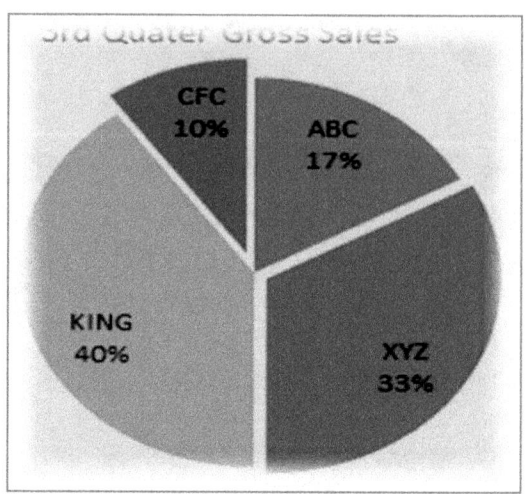

3rd Quarter Gross Sales Receipts

in percentage of 3 restaurants

The question is asking CFC sales of 3rd quarter from the graphs reported. It depicts 10% shown above.

The CFC restaurant earned 10% of 9,100,000 million, which is (9,100,000 X 0.01) = $910,000

TABLE CHART

What is table Chart?

Table contains data that known as table chart.

Typical Table chart

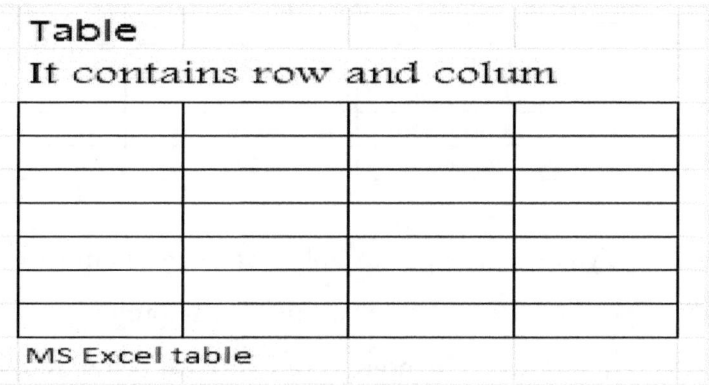

A Table presents the data as a table of rows and columns, and is used to see details and compare values.

The CBEST math questions derive information from graphs, charts, and tables' interpretations and computations.

CBEST Math Section:
California Basic Educational Skills Test

Sample table contain data		
Hourly sales for the day from 8:00 AM to 4:00 PM		
Hour	Hamburger	Salad
10:00 AM	92	120
11:00 AM	80	85
12:00 PM	70	77
1:00 PM	71	74
2:00 PM	72	75
3:00 AM	88	91
4:00 AM	112	111

Typical sales statistic table that creates number of questions, one question can be two products hourly trends. Second question can be hourly combined sales and third question can be how many more salad sold than hamburger etc.

1. From the table, we see that it creates Sales trend hourly since shop opens at 10:00. For instance, sales go up both products from 12:00 noon to 4:00 pm almost steady. However, exception is first two hours 10:00 and 11:00. CBEST question may ask like this from the above analysis.

CBEST Math Section:
California Basic Educational Skills Test

Prediction

Question statement 1:

If the sale pattern continues as per information provided above table, what do you predict sales pattern?

As per data provided, we can conclude that future business trends & growths will be positive manner of both products.

Comparison

Question statement 2:

Another question can be seen CBEST math based on similar table, how many hours (from 10:00 PM to 4:00 PM) more salad sold than hamburgers.

This question tells us to <u>compare</u> two sales food items. As we look at the table data, it shown salad sales are **more** than hamburger all the hours; however, one exception is 4:00 PM two products sale difference is (112-111)= 1, which is not considered more; therefore, its response correct response is 6 hours from 10:00 am to 5:00 pm.

Computation

Question statement 3:

Third category question can be computation types. One such question could be on which hour were the **most** sold of both products.

Look at the table chart, 4:00 P.M. is time when both items were sold most.

Chapter conclusion is that in math solution cases, some keywords are very special meaning like more, most, approximate, rounded off because they indicate mathematically much difference meaning than literally meaning. For example, average, mean, approximate, pattern, trend, and especially most and more are very much different meaning indeed. They will be discussed throughout the CBEST math section.

CBEST Math Section:
California Basic Educational Skills Test

Blank Page

CBEST Math Section:
California Basic Educational Skills Test

	Geometry	
	Geometry Triangle	255
	Triangle Shape	256
	Angle Measure Degree	259
	Pythagorean Theorem	260
	Angle within the circle	262
	Cube	264
	Square	264
	Rectangle	265
	Parallel Lines	267
	Parallelogram	268

CBEST Math Section:
California Basic Educational Skills Test

Geometric Various Triangle

What is triangle?

A trianlge makes three sides or arms that each arm end meets with other arm's end that makes a triangle, which creates 3 angles. These 3 angles sum always 180°.

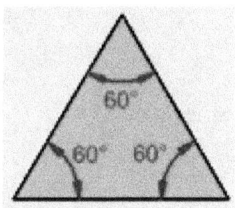

A Typical Equilateral Trianlge

Triangle Types are:

1. Equilateral: All three angles are 60°, therefore, all arms are equal.

2. Isosceles: Two arms are equal, therefore, that two sides opposite angles are equal, but angles' degree varies.

3. Scalene: Three arms are different length, thus, angles are various angles depending on arm's length

4. Acute Triangle: Angles are less than 90°

5. Right Triangle: One angle must have 90°

6. Obtuse Triangle: One angle must have more than 90°

Triangles' Shape :

Scalene Traingle:

3 angles are not equal, therefore, 3 sides are not same length.

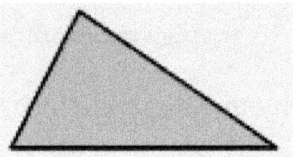

Isosceles Triangle:

Two sides are equal lenght, therefore, that two sides opposite angles are equal

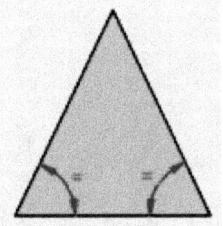

Equilateral Triangle:

Three angles are equal as a reuslt each angles measured 60°, which make all its sides lenght equal.

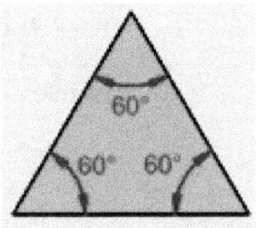

CBEST Math Section:
California Basic Educational Skills Test

Acute Triangle:

Angles are less than 90°

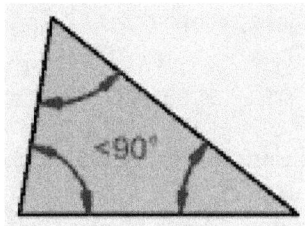

Right Triangle:

One angle must have 90°

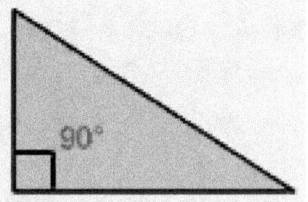

Obtuse Triangle:

One angle must have more than 90

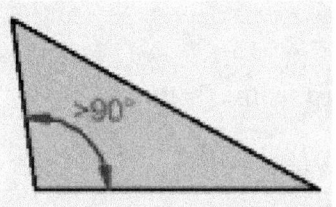

Triangle Area:

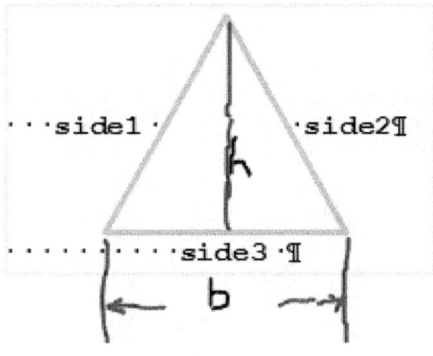

CBEST Math Section:
California Basic Educational Skills Test

Triangle Perimeter = Side1 + Side2 + side3

Triangle Area = $\frac{1}{2}$ b x h

Where b= base, h=height, x = multiplication sign

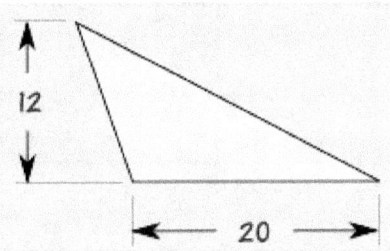

Area A= ½ (b * h) = ½ (20*12)=1/2 (240)=120

Learn from Examples:

What is the base of a triangle in which has base =6" and area=30sq".

    A = 30 , b =6, h=?

    A=1/2 (b*h) (replace given value into the fomula)

    or 30 = ½(6 *h)

    or 30 = 3h

    or 3h = 30

or h = 30/3 = 10"

It is important to remember that to figure out an area of a triangle, which needs to know the formula, not the angles' measure.

Angle:

What is angle?

Angle measures in degree between two intersecting lines where they meet

/b = 30°

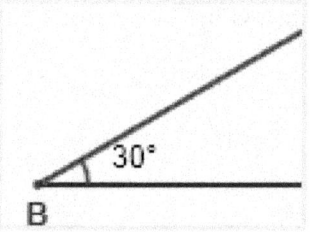

Typical angle shown above is 30°

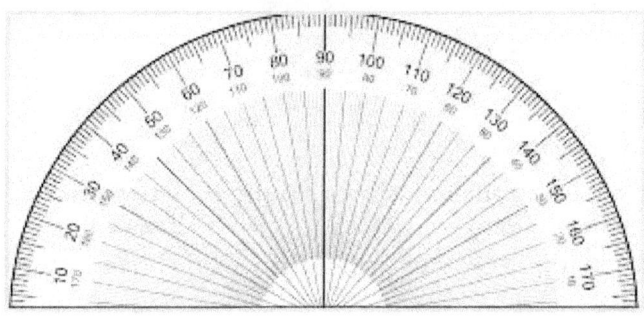

Typical Degree Measurement Tool

Pythagorean Theorem:

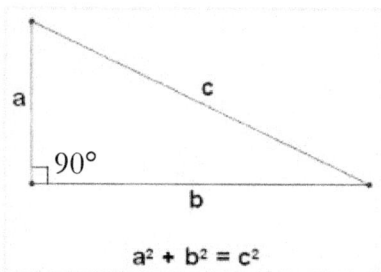

Pythagorean Theoream, a triangle must have one angle 90° i.e. Right Triangle.

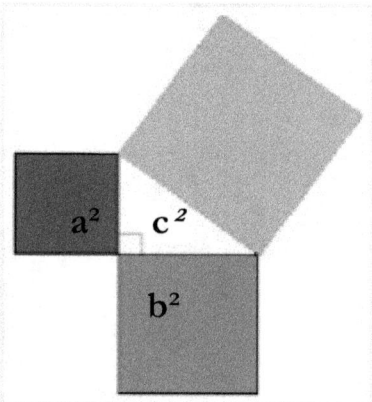

Pythagorean theoream based on Right Triangle (90°). It states that $c^2 = a^2 + b^2$, which reflected above picture.

CBEST Math Section:
California Basic Educational Skills Test

Learn from examples:

Example 1:

$a=4; b=5, c=?$

Replace value on this formula: $c^2 = a^2 + b^2$

or $a^2 + b^2 = c^2$

or $4^2 + 5^2 = c^2$

or $16 + 25 = 41$

$c^2 = 41$

Example 2:

What is vlue of a,

where $c=5, b=4$,

Plug in value on formula that gives the length of a.

$a^2 + b^2 = c^2$

or $a^2 + 4^2 = 5^2$

or $a^2 + 16 = 25$

or $a^2 = 25 - 16$ (similar term apply)

or $a^2 = 9$

or $a^2 = 3*3$

or $a^2 = 3^2$ (remove squre both sides)

or $a = 3$

Angle resides inside of a Circle:

There are 360° of arc in every circle no matter how big or small cirle.

Circle contains lines, segment, and angles, sector, arc and many more components.

An exhibit of circle components

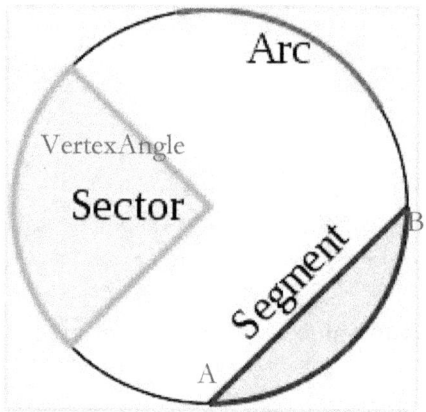

Arc:

An arc is a portion of the circumference of a circle. An arc is shown red colored.

Sector:

A circular sector is the portion of a circle that enclosed by two radii and an arc.

Chord : A line that links two points on a circle's arc

Segment:

A line that intersects a circle arc at two places.

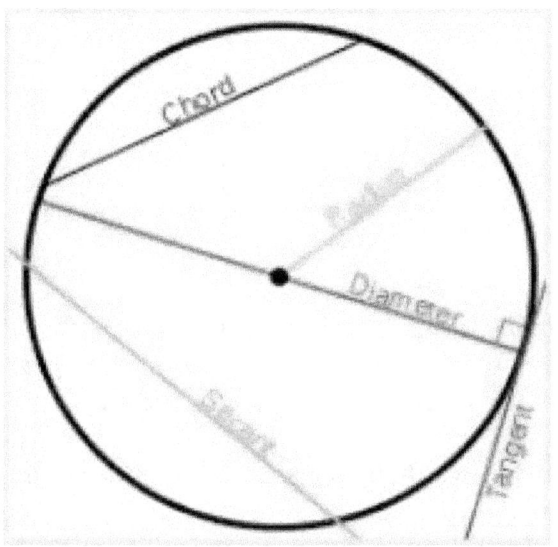

Diameter: A straight line that passes the center point of a circle and divided the circle into parts.

Radius: A radius is line shown above.

Tangent: A tangent is a line that touches a arc surface at one place/point only.

Secant: A straight line that cuts a arc in two places

Cube

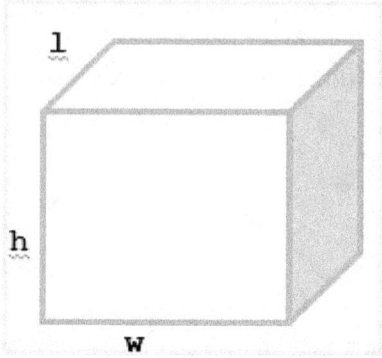

Volume = (l x w x h) = 3w or 3h or 3*l*

Since, cube has all sides are equal thus   3w=3h=3*l*

Surface Area = 2(l x w) + 2(l x h) +2(w x h)

l=length, w=width, h= height, x=multiplying sign

Square

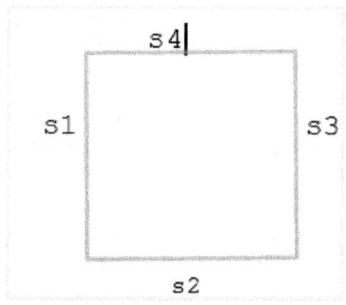

Square Perimeter = 4S (side multiply by 4)

Area = side1 x side2 or side squared

CBEST Math Section:
California Basic Educational Skills Test

Rectangle

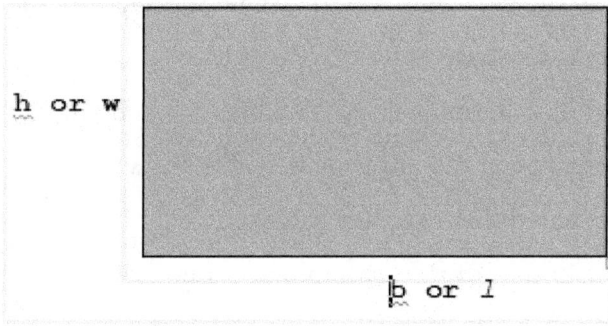

Perimeter = 2(b + h) or 2(l + w)
Area = (b x h) or l x w
b=base, h= height, l=length, w=width, x=multiplication symbol.

Learn from Examples:

> Length =4X – 6
> Perimeter = 60 feet
> Width =X
> <---What is the Length?--->

Alternately this problem can be presented CBEST test like this, no illustration provided:

What is the width?

Where a rectangle perimeter is 60 feet and length is 4 times of its width minus 6.

The problem above is algebric in nature in which test taker needs to know interpretation of word solving issue like time means multiplication i.e. 4X, subtraction like, – 6 etc.

CBEST Math Section:
California Basic Educational Skills Test

What is lenght of the rectangle?

Plug-in the value into the rectangle formula.

Rectangle Formula, $P = 2W + 2L$

where P = Peripheral, X =Width, L= Lenght.

Again, the problem did not provide width, however, it indirecctly stated that its width is X.

Length is, 4 times of its width minus 6. Let's say, width is X. Thus, when width is X, then its L or length is $(4X - 6)$.

Therefore,

$P = 2X + 2(4X - 6)$

$60 = 2X + 8X - 6X$

$60 = 10X - 6X$

$60 = 4X$

$4X = 60$

$X = 60/4$

$X = 15$

Rectangular Solid

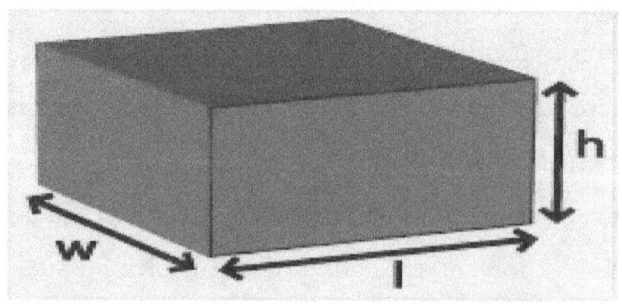

Volume= (l x w x h)

Surface=2(l x w) + 2(l x h) + 2(w x h)

Where l= length, h= height, w = width

Parrallel Lines:

Geometric parallel lines run in a plane straight two lines (side by side) that remain the same distance apart, which never intersects over their entire length/distance.

Line 1 = a ; Line 2 = b

   a || b

Transversal line = cd

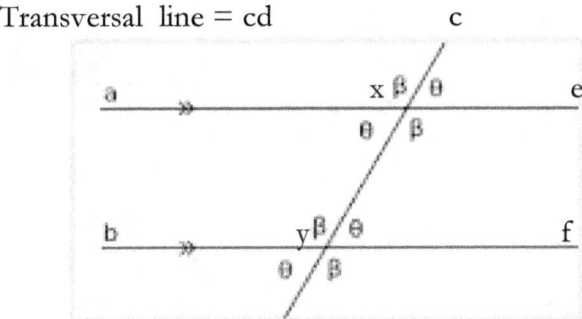

Line cd cuts parallel lines a and b, which create 8 anglels.

All angles that θ represnets measure equal

Similarly all angles β represents measure equal

A transversal line cd creates several angles that are decribed hereafter.

Angle ‹cxe & ‹axd = veritical angles

Angle cxe & ‹cyf = corresponding angles

Angle ‹axd & ‹cyf = alternate interrior angles

Angle ‹cxe & ‹byd = alternate enterior angles

Angle ‹cxa & ‹byd are called supplementary angles, which are transversal exterior on the same side.

- **Parallelogram**

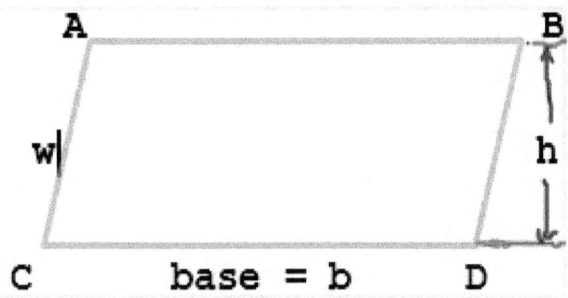

AB = CD; AC = BD, AB||CD, AC||BD

Area= (b x h)

Perimeter = 2(l + w)

b = base, h = height, w = wide

CBEST Math Section:
California Basic Educational Skills Test

ABOUT THE AUTHOR

Author is a teacher/instructor/writer of several academic books and professional development. He earned his B.S. & M.A. Edu from www.APU.Edu., CA. He has more than 25 years professional experience in IT management and teaching education. He also obtained adult teaching credentialing issued by the State of California.

www.ingramcontent.com/pod-product-compliance
Lightning Source LLC
Chambersburg PA
CBHW051634170526
45167CB00001B/184